Evelina Colacino, Guido Ennas, Ivan Halasz, Andrea Porcheddu,
Alessandra Scano (Eds.)
Mechanochemistry

Also of interest

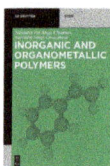

Inorganic and Organometallic Polymers
Pal Singh Chauhan, Singh Chundawat, 2019
ISBN 978-1-5015-1866-9, e-ISBN 978-1-5015-1460-9

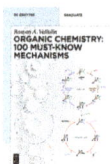

Organic Chemistry: 100 Must-Know Mechanisms
Valiulin, 2020
ISBN 978-3-11-060830-4, e-ISBN 978-3-11-060837-3

Advanced Composites
Davim (Ed.)
ISSN 2192-8983

Pure and Applied Chemistry.
The Scientific Journal of IUPAC
Burrows, Stohner (Eds.)
ISSN 0033-4545, e-ISSN 1365-3075

Mechanochemistry

A Practical Introduction from Soft to Hard Materials

Edited by
Evelina Colacino, Guido Ennas, Ivan Halasz,
Andrea Porcheddu and Alessandra Scano

DE GRUYTER

Editors

Prof. Evelina Colacino
Université de Montpellier
Institut Charles Gerhardt de Montpellier
(ICGM) – UMR 5253 CNRS, ENSCM, UM
8, Rue de l'Ecole Normale
340296 Montpellier
France
evelina.colacino@umontpellier.fr

Prof. Guido Ennas
Dipartimento di Scienze Chimiche e Geologiche
Università di Cagliari
Cittadella Universitaria
09042 Monserrato (Cagliari)
Italy
ennas@unica.it

Dr. Ivan Halasz
Division of Phys. Chem.
Ruđer Bošković Institute
Bijenička cesta 54
HR-10000 Zagreb
Croatia
Ivan.Halasz@irb.hr

Prof. Andrea Porcheddu
Dipartimento di Scienze Chimiche e Geologiche
Università di Cagliari
Cittadella Universitaria
09042 Monserrato (Cagliari)
Italy
porcheddu@unica.it

Dr. Alessandra Scano
Dipartimento di Scienze Chimiche e Geologiche
Università di Cagliari
Cittadella Universitaria
09042 Monserrato (Cagliari)
Italy
alescano80@tiscali.it

ISBN 978-3-11-060964-6
e-ISBN (PDF) 978-3-11-060833-5
e-ISBN (EPUB) 978-3-11-060847-2

Library of Congress Control Number: 2020948699

Bibliographic information published by the Deutsche Nationalbibliothek
The Deutsche Nationalbibliothek lists this publication in the Deutsche Nationalbibliografie;
detailed bibliographic data are available on the Internet at http://dnb.dnb.de.

© 2021 Walter de Gruyter GmbH, Berlin/Boston
Cover illustration: Prof. Dr. – Ing. Christoph Hartl, with the use of photographic material
and license from Getty Images Deutschland GmbH
Typesetting: Integra Software Services Pvt. Ltd.
Printing and binding: CPI books GmbH, Leck

www.degruyter.com

Preface

Mechanical processing to make new materials and substances is an ancient technique, with creation of new matter by grinding, rubbing, and different types of shear most likely being among the oldest practices known to humankind. Mechanically induced transformations that involve chemical change have been documented at least as long as 2000 years ago, for example in the method for winning mercury from cinnabarite outlined in Theophrastus' *Book of Stones*. Grinding in some form or other has been instrumental for the development of early medical remedies, something that we are often still reminded of by the mortar-and-pestle symbol found outside of many pharmacists' stores. Chemistry resulting from mechanical scratching and milling has attracted the attention of some of the legends of early chemistry and physical sciences, as seen in M. Faraday's explorations of reactivity of hydrated inorganic solids, or in the reported use of *Kügelchen* by *Wöhler*. While chemical reactions induced by mechanical action were established and well recognized in the early 20th century, with the subject given the name *mechanochemistry* in Ostwald's *The Fundamental Principles of Chemistry*, this area subsequently fell into disregard and oblivion. That is, until now – modern days have seen a true renaissance of mechanical reactivity across probably all major areas of chemical synthesis and materials chemistry, ranging from organic synthesis to nanoparticle materials.

This exciting spirit of new exploration and recognizing new opportunities offered by the cleaner, safer and simply "greener" solvent-free reaction medium of mechanochemistry is perfectly captured by this introductory textbook to the topic, edited by Colacino, Ennas, Halasz, Porcheddu and Scano. These authors are, each in their own way, pushing forward the new mechanochemistry which has been undergoing virtually explosive, previously unseen level of development. This development has created a clear need for a broader, but still accessible, overview of the field, in the form that could be attractive, approachable and taken as an introductory material to the wide range of applications that are now being addressed by mechanosynthesis by milling or grinding. This book covers that need in the most satisfactory way, providing an easy to follow overview of the field, which includes a brief and broad discussion of the area in Chapter 1, followed by a succinct, clear and useful discussion of the available apparatus in Chapter 2, highlighting their fundamental designs, and importantly the similarities and differences, advantages and disadvantages. The book also provides a more detailed focus on selected areas of mechanochemistry, some of which are based on traditional divisions of reactivity in inorganic systems (Chapter 3), while others provide a different line of reasoning. This is the case in Chapter 4, which provides an overview of mechanochemical organic reactivity, but presented by Colacino and Porcheddu through the lens of a set of experiments that can be implemented in teaching undergraduate curricula. This chapter indeed shows instructors how mechanochemistry can be used to teach

https://doi.org/10.1515/9783110608335-202

students to "think chemistry differently". Chapter 5 by Halasz offers an exciting overview of a relatively new topic of cocrystallization, in that way providing a taste of supramolecular chemistry to the book, while at the same time providing a brief glimpse into the evolving world of real-time studies of mechanisms that underlie mechanochemical reactions in a mill. An overview of coordination polymer mechanochemistry, which follows in Chapter 6 provided by Ennas and Scano is a logical extension from that topic, which gives a short overview how mechanochemical milling is helping push forward this new area of materials science. While providing an overview of modern mechanochemical synthesis is a challenge, it is even more so when the goal is to create a succinct, exciting, and well-organized textbook.

The authors have surmounted this challenge with outstanding success and I warmly recommend *Mechanochemistry – A Practical Introduction From Soft to Hard Materials* as an excellent introductory synthesis-by-mechanochemistry text to students, educators, afficionados and experts alike.

Tomislav Friščić, Prof. (McGill University, Montréal)

Acknowledgements

This book is based upon work from COST Action CA18112, supported by COST (European Cooperation in Science and Technology).

COST (European Cooperation in Science and Technology) is a funding agency for research and innovation networks. COST Actions help connect research initiatives across Europe and enable scientists to grow their ideas by sharing them with their peers. This boosts their research, career and innovation.

http://www.cost.eu/

Funded by the Horizon 2020 Framework Programme of the European Union

COST Action CA18112 – Mechanochemistry for Sustainable Industry (MechSustInd, www.mechsustind.eu) is a European Union networking initiative aimed to nurture and catalyse interactions between European and overseas researchers in Mechanochemistry.

COST Action CA18112 is supported for 4 years (2019-2023) by European COST funds and gather more than 130 scientists from 33 Countries within Europe. The international dimension of COST Action CA18112 goes out the borders of Europe and counts members from Canada, China, Mexico, Russia, Singapore and USA.

COST Action CA18112 aims to nucleate a critical mass of actors from EU research Institutions, enterprises and industries, bringing together different areas of expertise and application. This in turn will create the necessary synergy to establishing a vigorous multi-disciplinary network of European scientists, engineers, technologists, entrepreneurs, industrialists and investors addressing the exploitation of mechanical activation in the production of chemicals through sustainable and economically convenient practices on the medium and large scales. Among the objectives, there is the implementation of mechanochemistry as a common practice at Undergraduate Level, as a way to contribute to the development of the 'sustainable thinking' of the future generations of scientists. For more information www.mechsustind.eu

https://doi.org/10.1515/9783110608335-203

Authors would like to thank Prof. Christoph Hartl (Technische Hochschule, Köln, Germany), for the design and realization of the book cover art.

Authors are extremely grateful to Prof. Tomislav Friščić (Department of Chemistry, McGill University, Canada) for the endorsement of this book.

Contents

Editors' Biographical Sketches

Evelina Colacino received her double PhD (with European Label) in 2002 at the University of Montpellier II, France, and at the University of Calabria, Italy. She was appointed as research fellow at the Catholic University of Louvain (Belgium, 2003), working on the preparation of new hydantoin scaffolds as antibacterial agents, research scientist at Sigma-Tau Pharmaceuticals (Italy, 2004) and postdoctoral fellow at the University of Montpellier II (France, until 2007). She was hired as assistant professor in 2008 and associate professor of organic and green chemistry, since 2013, at the University of Montpellier, France. Her main research activities concern the development of eco-friendly methodologies for the preparation of biomolecules, heterocyclic compounds and hybrid materials by mechanochemistry (dry or wet grinding), with main focus on hydantoin scaffold and active pharmaceutical ingredients (*medicinal mechanochemistry*). She also investigates sustainable approaches to homogeneous or heterogeneous metal-catalyzed processes by combining enabling technologies (ultrasounds, microwaves and flow systems) with nonconventional media (e.g., glycerol, water, poly(ethylene)glycols and PEG-based ionic liquids) or in micellar conditions (in water and glycerol). She has (co)authored five book chapters, two patents and 70 peer-reviewed scientific publications.

She is a member of the American Chemical Society, the International Mechanochemical Association and Beyond Benign association for green chemistry education. She is a promoter of sustainability in higher education by integrating green chemistry at undergraduate level in organic chemistry courses, teaching laboratories and across the subdisciplines of chemistry, with a special focus on the fundamentals and the practice of mechanochemistry.

Since 2019, she leads the European Programme COST Action CA18112 (MechSustInd, 2019–2023) – 'Mechanochemistry for Sustainable Industry (www.mechsustind.eu and https://www.cost.eu).

Guido Ennas, graduated in chemistry at University of Cagliari (1986), received his master's degree in materials science at the Polytechnic of Turin (1988), and his PhD in chemistry at the University of Cagliari (1991). He worked as lecturer at the University of Cagliari from 1992 to 2001. Since 2001, he has been associate professor of inorganic chemistry at the Pharmacy Faculty of the University of Cagliari. His roles have included coordinator of the School of Pharmacy at the University of Cagliari and member of the Executive Board of the Consortium INSTM (National Interuniversity Consortium of Materials Science and Technology). Prof. Ennas is responsible of the Micro- and Nanostructured Materials Research Group, having more than 20-year research experience in design, synthesis and characterization of nanostructured, micro- and mesoporous materials, with particular relevance to the development of different inorganic materials and coordination polymers by mechanochemical approach. He has published more than 100 scientific articles and two national patents and presented his results at national and international conferences. He is reviewer for several international journals in the field of materials science.

https://doi.org/10.1515/9783110608335-205

Ivan Halasz graduated in chemistry in 2003 and obtained his PhD with Prof. Hrvoj Vančik in 2008 from the University of Zagreb. Following a two-year postdoctorate with Prof. Robert Dinnebier at the Max-Planck-Institute for solid-state research in Stuttgart, he joined as a faculty at the Ruđer Bošković Institute in Zagreb in 2012, where he holds the position of a senior researcher (equivalent to an associate professor) since 2016. His research interests are focused on solid state, including solid-state reactivity, mechanochemistry, powder X-ray diffraction, structural characterization and reaction mechanisms. He published around 80 research papers and coauthored two high-school chemistry textbooks. In 2016, he received the Croatian State Science Award for his contributions in developing in situ methods for the study of mechanochemical reactions. He is currently the leader of the working package of the COST Action CA18112 "Mechanochemistry for Sustainable Industry" (MechSustInd) dedicated to in situ monitoring.

Andrea Porcheddu studied chemistry at the University of Sassari and got his "Laurea" degree in chemistry in 1995 with first-class honors. His diploma thesis dealt with the synthesis of piperazic acid derivatives as CPP analogues, which was carried out under the direction of Dr. Massimo Falorni. He then undertook doctoral research, designing novel strategies to use cyanuric chloride (TCT) in friendly organic processes under the supervision of Prof. Maurizio Taddei at the University of Sassari and was awarded his PhD in 1999. He completed postdoctoral studies (2000) in the group of Prof. Charles Mioskowsky at the Louis Pasteur University (Strasbourg, France) working on the synthesis of quinuclidine derivatives. In 2001, he moved back to Sassari University where he was appointed as assistant professor. In January 2015, he joined the Chemistry Department of the University of Cagliari (Italy), where he currently has a permanent position as associate professor. His diverse experience ranges from the synthesis of chimera molecules possessing complex molecular architecture to environmentally friendlier alternatives for synthesis using the most advanced technologies such as solid-phase synthesis, combinatorial chemistry, green chemistry in eutectic solvents, microwaves and mechanochemical mixing. The second field of interest concerns the development of novel *Borrowing Hydrogen* and *Transfer Hydrogenation* strategies for making C–N bonds using tertiary amines or alcohols instead of more labile aldehydes. The scientific interests of Prof. Porcheddu are also focused on finding novel and highly efficient nonconventional catalysts (mainly Fe and Cu) and reagents for C–H bond activation, intending to minimize both waste production and energy consumption. Currently, his main scientific interest covers the field of ball-milling chemistry, and is devoted to the study of nonconventional transformations over the field of organic chemistry, which can reduce solvents, by-products and wastes, and they can be defined as environmentally sustainable.

Prof. Porcheddu has authored six-chapter books and he is author or coauthor of more than 90 scientific publications on refereed international journals (H index, Scopus = 32).

Presently, he represents Italy in the Management Committee of the European Programme COST Action CA18112 (MechSustInd) – "Mechanochemistry for Sustainable Industry" (MechSustInd).

Dr. Alessandra Scano graduated with Master of Science in pharmaceutical chemistry and technology (2005) and in cellular and molecular biology (2019) at the University of Cagliari. In 2009, she received her PhD in chemical science. She spent one year (2009) as postdoctorate at Materials and Surface Science Institute, University of Limerick (Ireland). From 2010 to 2013, she worked as researcher in synthesis (top-down and bottom-up approaches) and characterization of nanomaterials at the Asociacion de la Industria Navarra, Spain. Since 2014, she is back to University of Cagliari where she is part of the Micro- and Nanostructured Materials Research Group. She works on the development of micro- and nanostructured materials for biomedical applications with particular interest on the mechanochemical approach. She is author/coauthor of several papers published in international journals, and she has presented her results at several international conferences. Dr. Scano has also experience in design and management of international and national projects, and she is part of the EUPF (Register of Euro-Projects Designers and Managers – Europe Project Forum Foundation).

Guido Ennas, Alessandra Scano, Andrea Porcheddu, Ivan Halasz,
Evelina Colacino

1 Mechanochemistry: an overview and a historical account

Introduction

This chapter aims to give a brief overview of the historical development of mechanochemistry, a scientific field that is nowadays becoming very active and promising in materials science and powder technology,[1–2] and also in processing of organic compounds[3–4] and pharmaceutical formulations.[5]

A description of the most common milling equipment available in academic laboratories for the purposes of research and teaching is highlighted in Chapter 2. Chapter 3 introduces readers to inorganic mechanochemistry, while Chapter 4 reports examples of organic mechanochemical reactions. Examples of mechanochemistry of cocrystals and coordination polymers are highlighted in Chapters 5 and 6, respectively. Topics related to fundamental studies and advanced mechanochemical applications are outside the scope of this book and are not covered here. Interested readers are invited to refer to a more specialized and flourishing literature available in the field.

The term mechanochemistry refers to any chemical reaction that is mechanically induced. According to the International Union of Pure and Applied Chemistry (IUPAC), mechanochemical reaction is defined as *Chemical reaction that is induced by the direct absorption of mechanical energy.*[5–6]

More specifically, in the case of solid-state processes, this definition is frequently associated with reactions initiated by any type of mechanical treatment or involving reagents that were pretreated and activated mechanically,[4] to create active sites for chemical reactivity, or increase the active surface of substances that can then more efficiently coalesce and react. Known since millennia,[1] mechanochemistry has been recently acknowledged by the IUPAC as one of the top 10 world-changing emerging technologies to enhance chemical synthesis.[7] The growing importance of mechanochemistry has also been recognized by prestigious funding organizations for research and innovation networks in Europe. Indeed, the European Cooperation in Science and Technology (COST)[8] association funded in 2019 the COST Action CA18112 *Mechanochemistry for Sustainable Industry* (MechSustInd),[9–11] aiming to foster technological and scientific growth of mechanochemistry across Europe, encompassing academic and industrial partners across 33 European countries and beyond.

Figure 1.1[12] schematizes the development of mechanochemistry, which is chronologically divided into four stages[1]: (1) *inadvertent mechanochemistry*, dating back to the prehistoric time; (2) its recognition as a technique to promote solid-state

https://doi.org/10.1515/9783110608335-001

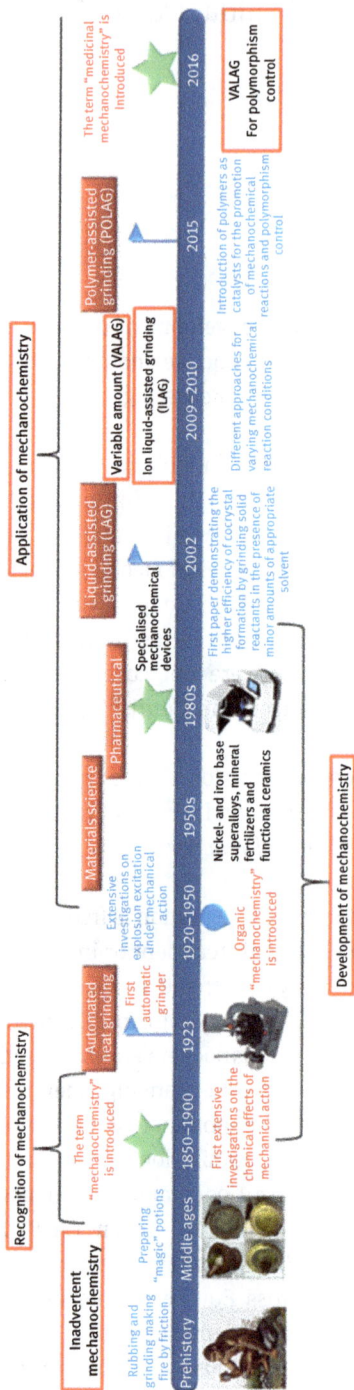

Figure 1.1: "Ages" of mechanochemistry. Image reproduced from reference[12] with the permission of Elsevier.

reactions, occurring during the middle of the nineteenth century; (3) its early development mainly in the field of material science, which includes both the first pioneering investigations on the chemical effects due to mechanical action and the construction of automatic grinders; and (4) the *modern period* characterized by its extensive application also to new areas of investigations (e.g., pharmaceutical formulation and "medicinal mechanochemistry"[13]) and the development of advanced mechanochemical processes involving the use of additives for polymorphism and reactivity control.[14–15]

The beginning of *inadvertent mechanochemistry* dates back into prehistory, where rubbing and grinding were used to prepare foodstuff, to make fire by friction and later to treat minerals, paints and medicines. The earliest known document related to mechanochemistry is found in the book *On Stone* from the year 315 BC written by Theophrastus of Eresus (who was a student of Aristotle), wherein the reduction of cinnabar (HgS) to mercury by grinding in a copper mortar and pestle is reported. Grinding and its traditional instrument, the mortar and pestle, can be regarded as "the first engineering technology."[1] Despite this earlier discovery, the first systematic studies of mechanochemical reactions were carried out only in the nineteenth century. In 1820, Michael Faraday described the reduction of silver chloride by grinding with zinc, tin, iron and copper in a mortar, and this method was called the "dry way" of inducing reactions.[16] He reported that silver chloride reacted with zinc in a fast and highly exothermic reaction. In 1866, Matthew Carey Lea investigated mechanical action to induce a chemical response.[17] The most remarkable finding of Lea was related to the decomposition of mercuric and silver chlorides. Both these compounds decomposed while triturated in a mortar, although they were known to melt or sublime upon heating.[18] In the same period, Walthère Spring carried out a large-scale systematic study of the mechanical action effects on chemical processes, which anticipated further investigations.[19] The main motivation for research endeavors of Spring and Lea was to understand the fundamental nature of chemical reactions under pressure and shear. Grinding, sliding and other forms of mechanical action are important components of many technological processes and they are often accompanied by (unwanted) chemical changes.[1]

During the first half of the twentieth century, mechanochemistry went through a slow growth period while its practical potential was recognized, and it became a source of motivation for further research.[1] F. M. Flavitsky, interested in solid-state reactions induced by grinding and in qualitative chemical analysis, investigated solid chemicals that could be used to identify, in a fast and accurate way, several cations and anions by simply rubbing small amounts of the unknown substance with an appropriate sequence of reactants.[20] It is noteworthy that in 1919, Wilhelm Ostwald, in his textbook on general chemistry, included mechanochemistry as one of the subdisciplines of chemistry, together with thermochemistry, electrochemistry, sonochemistry and photochemistry.[21] During the Second World War, the development of mechanochemistry was an impulse due to increase of military interest on the mechanical initiation and sensitivity of explosives.[22] Studies related to this subject put the basis for the mechanical impact model assuming the formation of hot spots giving rise to ignition.[23] Later, such model was also extended to explain chemical reactions caused by sliding. However, this theory did not find favor with all scientists. In fact, another theory affirmed that mechanochemical reactivity was caused by the direct breaking of bonds due to the mechanical impact, as disclosed by Fink and Hofmann.[1, 24] In 1949, mechanochemistry found its first demonstration in a teaching laboratory when students at the University of Saint Petersburg performed qualitative analysis of inorganic mixtures using the Flavitsky's method.[1] In the late 1960s, the synthesis of complex oxide dispersion-strengthened alloys for high-temperature structural applications by ball milling was developed by John Benjamin and his coworkers at the International Nickel Company.[25] Benjamin suggested that it should be generally applicable to the preparation of metastable structures and with his coworkers demonstrated that mechanical alloying (MA) was suitable for the synthesis of solid solution alloys and otherwise immiscible systems, as well as metal composites and compounds.[26,27] In 1983, C. Koch and his group began the "solid-state amorphization" time.[28] The starting point for Koch was the results published by R. L. White[29] on the preparation of amorphous Nb_3Sn from powdered elemental niobium and tin. Koch also prepared amorphous $Ni_{60}Nb_{40}$ showing strong similarities between the mechanochemically prepared substance and the one obtained upon melt quenching. Since these studies, MA has become widely used for the synthesis of noncrystalline metals, alloys and intermetallic compounds,[30] ceramics, composites and nanocomposites by milling metal powders, also in reactive atmospheres.[24, 30–33]

Nowadays, mechanochemistry is gaining importance due to the need for development of cleaner manufacturing processes and is recognized as a potential alternative to traditional chemical syntheses, according to the principles of green chemistry,[34] responding to an increasing demand for clean processes and ecoconscious reaction conditions. Indeed, the milling process[35] is finding more and more space as a valid alternative to solution-based procedures in organic[4, 36–38] and organometallic syntheses,[39–41] enzymatic catalysis,[42] for the preparation of cocrystals,[43] metal-organic[44–45] and nano(biohybrid)materials,[2, 46] coordination polymers,[47] to access new pharmaceutical materials[12–13, 48–49] (inducing amorphization or promoting recrystallization of amorphous phases)[50] and opening new perspectives in biotechnology and biomedical fields for the preparation of advanced materials (*e.g.*, for the postsynthesis or *in situ* drug encapsulation, to obtain bioconjugates or for surface functionalization).[35, 51–54]

Over the last three decades, several studies have demonstrated the success of mechanochemistry not only to discover novel crystalline forms[13] or new drugs,[55] but also in processes involving the industrial formulation of pharmaceutical materials, leading to a better solubility of existing active pharmaceutical ingredients (APIs), increased dissolution rate, thermal and moisture stability and improved compressibility[56] upon mechanical treatment. Such properties are related not only to the crystal structure of the API but also to its particle size distribution and morphology, properties that can be targeted in a milling process[57] imparting to the system different modes of mechanical action (*e.g.*, impact, compression, shearing, stretching and grinding). These results have represented the beginning of a solid-state chemistry and pharmaceutical technology sodality.

It is also noteworthy to mention the recent development of new approaches of mechanochemistry such as kneading,[58–59] or twin-screw extrusion[60–63] for low-energy large-scale manufacturing of pharmaceutical materials (also having the advantage of avoiding high temperatures, sometimes responsible of APIs degradation), or using liquid-assisted grinding process,[53, 64–65] particularly successful in controlling crystallinity of the milling product and to prepare novel polymorphs, salts or cocrystals of APIs.

Enormous progress, especially in academic research, has already been accomplished. The next step is its implementation as a key strategy and a green and sustainable technology for real-world applications.

References

[1]   Takacs, L. The historical development of mechanochemistry. Chem. Soc. Rev. 2013, 42, 7649–7659, and references cited therein.
[2]   Baláž, P., Achimovičová, M., Baláž, M., Billik, P., Cherkezova-Zheleva, Z., Criado, J. M., Delogu, F., Dutková, E., Gaffet, E., Gotor Martinéz, F. J., Kumar, R., Mitov, I., Rojac, T., Senna, M., Streletskii, A., Wieczorek-Ciurowa, K. Hallmarks of mechanochemistry: from nanoparticles to technology. Chem. Soc. Rev. 2012, 42, 7571–7637.
[3]   Wang, G.-W. Mechanochemical organic synthesis. Chem. Soc. Rev. 2013, 42, 7668–7700.
[4]   James, S. L., Adams, C. J., Bolm, C., Braga, D., Collier, P., Friščić, T., Grepioni, F., Harris, K. D. M., Hyett, G., Jones, W., Krebs, A., Mack, J., Maini, L., Orpen, A. G., Parkin, I. P., Shearouse, W. C., Steed, J. W., Waddell, D. C. Mechanochemistry: opportunities for new and cleaner synthesis. Chem. Soc. Rev. 2012, 41, 413–447.
[5]   Boldyreva, E. V. Mechanochemistry of inorganic and organic systems: what is similar, what is different? Chem. Soc. Rev. 2013, 42, 7719–7738.
[6]   IUPAC. Compendium of Chemical Terminology. 2nd ed. (the "Gold Book"). Compiled by A. D. McNaught and A. Wilkinson. Blackwell Scientific Publications, Oxford, 1997.
[7]   Gomollón-Bel, F. Ten Chemical Innovations That Will Change Our World. Chem. Int. 2019, 41, 12–17.
[8]   For more information: http://www.cost.eu/.
[9]   For more information on COST Action CA18112 'Mechanochemistry for Sustainable Industry': http://www.mechsustind.eu/.
[10] Hernández, J. G., Halasz, I., Crawford, D. E., Krupička, M., Baláž, M., André, V., Vella-Zarb, L., Niidu, A., García, F., Maini, L., Colacino, E. European Research in Focus: Mechanochemistry for Sustainable Industry (COST Action MechSustInd). Eur. J. Org. Chem. 2020, 8–9.
[11] Baláž, M., Vella-Zarb, L., Hernandez, J. G., Halasz, B., Crawford, D. E., Krupička, M., André, V., Niidu, A., Garcia, F., Maini, L., Colacino, E. Mechanochemistry: a disruptive innovation for the industry of the future. Chem. Today 2019, 37, 32–34.
[12] Hasa, D., Jones, W. Screening for new pharmaceutical solid forms using mechanochemistry: a practical guide. Adv. Drug Deliv. Rev. 2017, 117, 147–161.

[13] Tan, D., Loots, L., Friščić, T. Towards medicinal mechanochemistry: evolution of milling from pharmaceutical solid form screening to the synthesis of active pharmaceutical ingredients (APIs). Chem. Commun. 2016, 52, 7760–7781.

[14] Hasa, D., Rauber, G. S., Voinovich, D., Jones, W. Cocrystal Formation through Mechanochemistry: from Neat and Liquid-Assisted Grinding to Polymer-Assisted Grinding. Angew. Chem. Int. Ed. 2015, 54, 7371–7375.

[15] Friščić, T., Reid, D. G., Halasz, I., Stein, R. S., Dinnebier, R. E., Duer, M. J. Ion- and Liquid-Assisted Grinding: improved mechanochemical synthesis of metal–organic frameworks reveals salt inclusion and anion templating. Angew. Chem. Int. Ed. 2010, 49, 712–715.

[16] Faraday, M., Shi, Q. J. The decomposition of chloride of silver by hydrogen and by zinc. Lit. Arts. 1820, 8, 374–376.

[17] Lea, M. C. Researches on the latent image. Br. J. Photo 1866, 13, 84.

[18] Lea, M. C. On endothermic decompositions obtained by pressure; Part II, Transformations of energy by shearing stress. Am. J. Sci. 1893, 276, 413–420.

[19] Spring, W. V.; Recherches sur la propriété que possèdent les corps de se souder sous l'action de la pression. Hayez, 1880.

[20] Flavitsky, F. M. Russ. Zh. Phyz. Khim. 1902, 8, 34.

[21] Ostwald, W. Die chemische Literatur und die Organisation der Wissenschaft. Akad. Verlag. Gesel. 1919, Vol. 1.

[22] Bowden, F. P., Yoffe, A. D. Initiation and Growth of Explosion in Liquids and Solids. CUP Archive, 1952.

[23] Gilman, J. During detonation chemistry may precede heat. Materials science and technology. Mater. Sci. Tech. 2006, 22, 430–437.

[24] Fink, M., Hofmann, U. Oxydation von Metallen unter dem Einfluss der Reibung. Z. Anorg. Allg. Chem. 1933, 210, 100–104.

[25] Benjamin, J. S. Dispersion strengthened superalloys by mechanical alloying. Metal. Trans. 1970, 1, 2943–2951.

[26] Benjamin, J. S., Bomford, M. J. Dispersion strengthened aluminum made by mechanical alloying. Metal. Trans. A 1977, 8, 1301–1305.

[27] Benjamin, J. S. New materials by mechanical alloying techniques. Arzt, E., Schultz, L., Eds. DGM Informationgesellschaft, Oberursel, Germany, 1989, 3–18.

[28] Koch, C., Cavin, O., McKamey, C., Scarbrough, J. Preparation of "amorphous" $Ni_{60}Nb_{40}$ by mechanical alloying. Appl. Phys. Lett. 1983, 43, 1017–1019.

[29] White, R. L. Ph.D. dissertation. Stanford University, 1979.

[30] Ennas, G., Magini, M., Padella, F., Pompa, F., Vittori, M. On the formation of Pd_3Si by mechanical alloying solid-state reaction. J. Non-Cryst. Solids 1989, 110, 69–73.

[31] Tabor, D. The Friction and Lubrication of Solids. Oxford University Press, 1964.

[32] Lynch, A., Rowland, C. The History of Grinding. Society for Mining, Metallurgy and Exploration. Inc.(SME), Littleton 2005. ISBN-13: 978–0873352383.

[33] Cocco, G., Enzo, S., Schiffini, L., Battezzati, L. X-ray diffraction study of the amorphization process by mechanical alloying of the Ni-Ti system. Mat. Sci. Eng. 1988, 97, 43–46.

[34] Erythropel, H. C., Zimmerman, J. B., de Winter, T. M., Petitjean, L., Melnikov, F., Lam, C. H., Lounsbury, A. W., Mellor, K. E., Janković, N. Z., Tu, Q., Pincus, L. N., Falinski, M. M., Shi, W., Coish, P., Plata, D. L., Anastas, P. T. The Green ChemisTREE: 20 years after taking root with the 12 principles. Green Chem 2018, 20, 1929–1961.

[35] Shakhtshneider, T. P., Boldyrev, V. V. Mechanochemical Synthesis and Mechanical Activation of Drugs. In: Reactivity of Molecular Solids, Boldyreva, E., Boldyrev, V., Ed. John Wiley & Sons, Chichester, UK, 1999, 271–312.

[36] Rodriguez, B., Bruckmann, A., Rantanen, T., Bolm, C. Solvent-free carbon-carbon bond formations in ball mills. Adv. Synth. Catal. 2007, 349, 2213–2233.

[37] Tan, D., Friščić, T. Mechanochemistry for organic chemists: an update. Eur. J. Org. Chem. 2018, 10, 18–33.

[38] Friščić, T., Mottillo, C., Titi, H. M. Mechanochemistry for Synthesis. Angew. Chem. Int. Ed. 2020, 59, 1018–1029.

[39] Bala, M. D., Coville, N. J. Organometallic chemistry in the melt phase. J. Organomet. Chem. 2007, 692, 709–730.

[40] Tan, D., Garcia, F. Main group mechanochemistry: from curiosity to established protocols. Chem. Soc. Rev. 2019, 48, 2267–2496.

[41] Do, J.-L., Tan, D., Friščić, T. Oxidative mechanochemistry: direct, room-temperature, solvent-free conversion of palladium and gold metals into soluble salts and coordination complexes. Angew .Chem. Int. Ed. 2018, 57, 2667 –2671, 2018, 2667–2671.

[42] Bolm, C., Hernandez, J. From synthesis of amino acids and peptides to enzymatic catalysis: a bottom-up approach in mechanochemistry. ChemSusChem. 2018, 11, 1410–1420.

[43] Braga, D., Maini, L., Grepioni, F. Mechanochemical preparation of co-crystals. Chem. Soc. Rev. 2013, 42, 7638–7648.

[44] Lazuen-Garay, A., Pichon, A., James, S. L. Solvent-free synthesis of metal complexes. Chem. Soc. Rev. 2007, 36, 846–855.

[45] Pilloni, M., Padella, F., Ennas, G., Lai, S., Bellusci, M., Rombi, E., Sini, F., Pentimalli, M., Delitala, C., Scano, A., Cabras, V., Ferino, I. Liquid-assisted mechanochemical synthesis of an iron carboxylate Metal Organic Framework and its evaluation in diesel fuel desulfurization. Micropor. Mesopor. Mat. 2015, 213, 14–21.

[46] Zhu, S. E., Li, F., Wang, G.-W. Mechanochemistry of fullerenes and related materials. Chem. Soc. Rev. 2013, 42, 7535–7570.

[47] Cabras, V., Pilloni, M., Scano, A., Lai, R., Aragoni, M. C., Coles, S. J., Ennas, G. Mechanochemical Reactivity of Square-Planar Nickel Complexes and Pyridyl-Based Spacers for the Solid-State Preparation of Coordination Polymers: The Case of Nickel Diethyldithiophosphate and 4,4′-Bipyridine. Eur. J. Inor. Chem. 2017, 13, 1908–1914.

[48] André, V., Quaresma, S., da Silva, J. L. F., Duarte, M. T. Exploring mechanochemistry to turn organic bio-relevant molecules into metal-organic frameworks: a short review. Beilstein J. Org. Chem. 2017, 13, 2416–2427.

[49] Quaresma, S., André, V., Fernandes, A., Duarte, M. T. Mechanochemistry – A green synthetic methodology leading to metallodrugs, metallopharmaceuticals and bio-inspired metal-organic frameworks. Inorg. Chim. Acta. 2017, 455, 309–318.

[50] Charnay, C., Porcheddu, A., Delogu, F., Colacino, E. New and up-and-coming perspectives for an unconventional chemistry: from molecular synthesis to hybrid materials by mechanochemistry. In: Green Synthetic Processes and Procedures, Edited by Ballini, R., Ed. RSC Green Chemistry Series (2019), Ch. 9.

[51] Rubio, N., Mei, K. C., Klippstein, R., Costa, P. M., Hodgins, N., Tzu-Wen Wang, J. T.-W., Festy, F., Abbate, V., Hider, R. C., Chan, K. L. A., Al-Jamal, K. T. Solvent-Free Click-Mechanochemistry for the Preparation of Cancer Cell Targeting Graphene Oxide. ACS Appl. Mater. Interfaces 2015, 7, 18920–18923.

[52] Mei, K.-C., Guo, Y., Bai, J., Costa, P. M., Kafa, H., Protti, A., Hider, R. C., Al-Jamal, K. T. Organic Solvent-Free, One-Step Engineering of Graphene-Based Magnetic-Responsive Hybrids Using Design of Experiment-Driven Mechanochemistry. ACS Appl. Mater. Interfaces 2015, 7, 14176–14181.

[53] Friščić, T., Trask, A. V., Jones, W., Motherwell, W. D. S. Screening for inclusion compounds and systematic construction of three-component solids by liquid-assisted grinding. Angew. Chem., Int. Ed. 2006, 45, 7546–7550.

[54] Pilloni, M., Ennas, G., Casu, M., Fadda, A. M., Frongia, F., Marongiu, F., Sanna, R., Scano, A., Vamenti, D., Sinico, C. Drug silica nanocomposite: preparation, characterization and skin permeation studies. Pharm. Dev. Technol. 2013, 18, 626–633.

[55] Braga, D., d'Agostino, S., Dichiarante, E., Maini, L., Grepioni, F. Dealing with Crystal Forms (The Kingdom of Serendip?). Chem. Asian J. 2011, 6, 2214–2223.

[56] Trask, A. V., Motherwell, W. D. S., Jones, W. Pharmaceutical Cocrystallization: Engineering a Remedy for Caffeine Hydration. Cryst. Growth Des. 2005, 5, 1013–1021.

[57] Carlier, L., Baron, M., Chamayou, A., Couarraze, G. Greener pharmacy using solvent-free synthesis: Investigation of the mechanism in the case of dibenzophenazine. Powder Technol. 2013, 240, 41–47.

[58] Braga, D., Giaffreda, S. L., Grepioni, F., Pettersen, A., Maini, L., Curzi, M., Polito, M. Mechanochemical preparation of molecular and supramolecular organometallic materials and coordination networks. Dalton Trans. 2006, 1249–1263.

[59] Braga, D., Grepioni, F., Maini, L. The growing world of crystal forms. Chem. Commun. 2010, 46, 6232–6242.

[60] Medina, C., Daurio, D., Nagapudi, K., Alvarez-Nunez, F. Manufacture of pharmaceutical co-crystals using twin screw extrusion: a solvent-less and scalable process. J. Pharm. Sci. 2010, 99, 1693–1696.

[61] Morrison, H., Fung, P., Tran, T., Horstman, E., Carra, E., Touba, S. Use of Twin Screw Extruders as a Process Chemistry Tool: Application of Mechanochemistry To Support Early Development Programs. Org. Process Res. Dev. 2018, 22, 1432–1440.

[62] Crawford, D., Casaban, J., Haydon, R., Giri, N., McNally, T., James, S. L. Synthesis by extrusion: continuous, large-scale preparation of MOFs using little or no solvent. Chem. Sci. 2015, 6, 1645–1649.

[63] Crawford, D.E, Porcheddu, A., McCalmont, A.S., Delogu, F., James, S.L., Colacino, E. Solvent-free, Continuous Synthesis of Hydrazone-Based Active Pharmaceutical Ingredients by Twin-Screw Extrusion. ACS Sustainable Chem. Eng. 2020, 8, 12230–12238.

[64] Friščić, T., Jones, W. Recent Advances in Understanding the Mechanism of Cocrystal Formation via Grinding. Cryst. Growth Des. 2009, 9, 1621–1637.

[65] For more details on liquid-assisted grinding (LAG) process see also Chapter 2.

Evelina Colacino, Andrea Porcheddu, Ivan Halasz,
Alessandra Scano, Guido Ennas

2 From manual grinding to automated ball mills

The search for the "universal solvent" (*alkahest*)[1] that fascinated the alchemists is still a persistent topic of the modern (sustainable) chemistry. Indeed, one of the possible approaches to redesign and innovate many classic organic reactions and processes is to make them ever more efficient from both the resource consumption and the economic point of view. Solvents (often also highly toxic) constitute up to 90% of the mass of the reaction system, and in addition to the related safety concerns, environmental impact of their widespread use is also a topic of public debate on the climate changes. In this regard, the way chemists think about chemistry constitutes a social responsibility and drives for a change, with the objective to propose innovative sustainable industrial processes and manufacturing.

The recent developments in the area of solid-state reactivity and the use of mechanochemistry as a tool to enable chemical synthesis through direct absorption of mechanical energy[2–3] contribute to the development of a "new way of doing chemistry," recently referred to as "Chemistry 2.0."[4] This approach is a viable alternative to solvent-based chemistry, rendering the current chemistry cleaner (leading to improved *waste economy*), energy efficient (*energy economy*),[5] achievable with *reagent economy* (usually a mechanochemical reaction can be performed using stoichiometric amounts of reactants) and *time economy* (improved kinetics are often observed and productivity increased per unit of time). Compared to solution-based procedures,[6, 7] mechanochemistry allows novel reactivities[8] with access to compounds otherwise impossible to obtain in solution (due to insolubility problems, poor reactivity due to steric hindrance, instability in solution or upon heating, etc.), with enhanced selectivity or presenting a selectivity switch.[9]

Mechanochemistry is an international and interdisciplinary field of investigation[10–12] at the interface of chemistry and mechanical engineering, encompassing organic, supramolecular, inorganic, organometallic and metal–organic chemistry, as witnessed by several dedicated reviews and books.[13–17] Cross-connected expertise merge to bring fundamental knowledge (to disclose the reaction mechanisms) and to advance the implementation of mechanochemistry at the industrial level, by characterizing the milling dynamics, which is, at the moment of publication of this book, investigated for only a very few cases. Indeed, the kinetics and thermodynamics of mechanochemical transformations at present remain poorly understood.[18–20]

The roots of mechanochemistry dates back to the Stone Age: various substances were smashed into powders (*e.g.*, leaves, seeds, wheat and clays) for medicinal purposes or for cooking using the mortar and pestle "technology," selecting smooth or coarse "ground stone" materials (basalt, rhyolite, granite, etc.) depending on the substances to be ground (Figure 2.1, panel a).

https://doi.org/10.1515/9783110608335-002

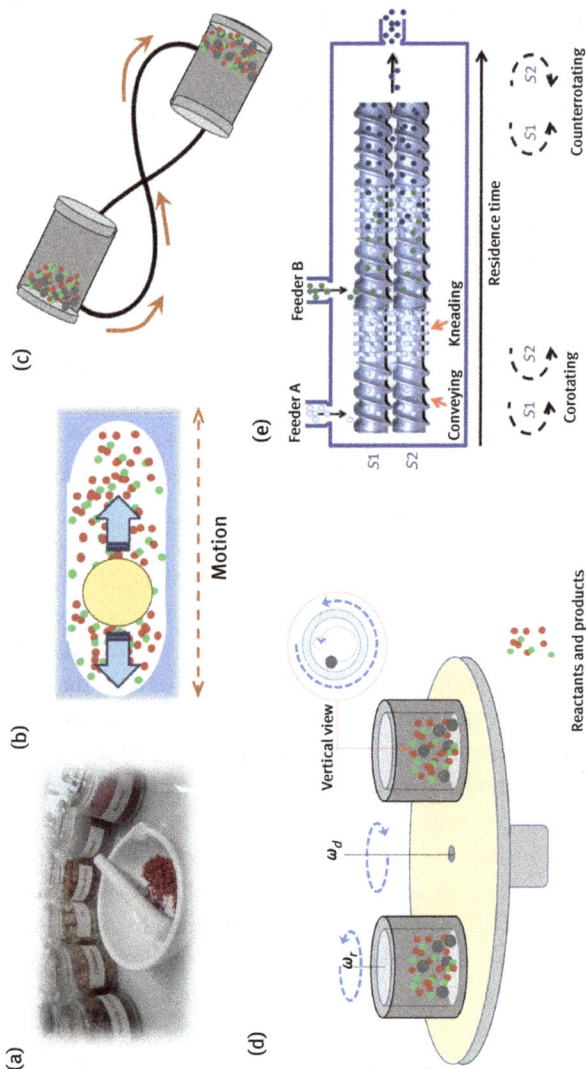

Figure 2.1: (a) Mortar and pestle; laboratory- and large-scale milling tools: (b) schematic cross section for horizontally oscillating vibratory mixer mill (for mixer mill with vertical oscillations, almost spherical grinding jars are mounted); (c) movement of the reaction chamber in a SPEX mixer mill in the shape of the number 8; (d) planetary ball mill and vertical view of the movement of the balls inside the jar; (e) schematic twin-screw extrusion (TSE) setup. Panel (e) is adapted from reference[21] with the permission of Royal Society of Chemistry.

Mortar and pestle is still an easy to use and a universal tool commonly found in undergraduate and research laboratories and it can be used to conduct "mechanochemically activated" reactions. However, the results obtained by manual grinding can be unreproducible, because of both human and environmental factors which are difficult to control such as (a) the manual force applied to the system, variable over time during processing and different for every operator, (b) different ambient humidity in the seasons and latitudes and (c) the (un)stability of some reactants to moisture or air. Additionally, safety concerns should also be considered, especially when handling toxic or volatile compounds. More controlled conditions, homogeneous and intense grinding, also for prolonged reaction times (i.e., several hours), are possible by using automatized grinders, that can also be cooled or heated and with adjustable exerted pressure.

Impact, compression, shearing, stretching and grinding are different ways to deliver mechanical energy into a system. As a result of the mechanical milling, particle size can be reduced, crystalline materials can become amorphous or undergo to a phase transition. The chemical reactivity is enhanced by the creation of active sites or fresh active surfaces, increased contacts between particles and possible coalescence before reacting.

More commonly, solvent-free laboratory-scale mechanochemical reactions are conducted using tools such as automated ball mills, differing essentially by the way in which the mechanical energy is imparted to the system (Figure 2.1, panels b–d). In general, they all involve a simple working principle: the reactants are charged in a closed reaction vessel (jar) containing balls. During the periodic motion, balls inside the reactor collide with each other at a high velocity, with the reactants and with the reactor walls, exerting a mechanical loading to the processed powder particles. The occurrence and the intensity of mechanical shocks depends on the process conditions (set by the operator) and by the specific action of the ball mill (different for a shaker or a mixer mill and a planetary mill) (Figure 2.1, panels b–d).

Attritors are also used to mechanically activate a reaction and usually are characterized by a relatively low energy, translating in frictional dynamics. However, depending on the operational conditions, impulsive dynamics can also be present. For this reason, a clear distinction between the two classes cannot be made.

In mixer (or shaker) mills, the reaction jars are rapidly shaken (horizontally or vertically) at a frequency up to 60 Hz, shearing and grinding the reactants together (Figure 2.1, panel b). Depending on the type of the milling device (and the jar material available) mixer mills can process sample quantities between 0.01 and 50 mL and are suitable for laboratory investigations up to a gram scale.

Mixer mills with a so-called "shape of 8" movement[22] also exists (Figure 2.1, panel c): the reactor undergoes simultaneously an angular harmonic displacement and rotation in the equatorial plane, with an operational frequency typically of 14.6 Hz.

Planetary mills display a mechanical action and working conditions quite different from mixer mills, with the jar movement compared to motion of planets around a sun. One or more reactors, mounted on a spinning disk (so-called *solar wheel*) rotating at an angular frequency ω_d, rotate on its own axis counterdirectionally at an angular frequency ω_r (Figure 2.1, panel d). The most popular planetary mills offer capacities between 12 and 500 mL, with rotational speeds up to 1100 rpm, resulting in centrifugal acceleration up to almost 100 times the Earth's gravity.

The throughput of mechanochemical reactions in batches can be highly increased using a planetary system with adapters holding multiple vessels (with capacities between 2 and 250 mL) and able to process up to 48 samples simultaneously (Figure 2.2). This approach, named as *parallel mechanochemistry*[23] was recently applied to the preparation of benzoxazine derivatives.

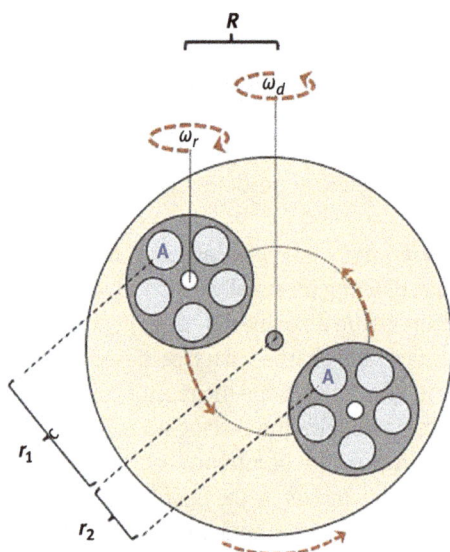

Figure 2.2: Schematic representation of a multisampling planetary mill endowed with *lunar movement*. Legend: R = distance between the jar axis and the disk center (solar wheel) axis; r_1 = distance between the vial A placed at the edge, and the solar wheel axis; r_2 = distance between the vial A placed toward the center (after 180° rotation), and the solar wheel axis.

As for a planetary ball mill (Figure 2.1, panel d) the adaptors (containing several vials) rotate at an angular frequency ω_d, and in the opposite direction at an angular frequency ω_r. The distance between the adaptors and the support disk is constant (R), but not in the case of the vial (A) (Figure 2.2). Indeed, during milling and over the time, the vials (A) are not always placed at the same distance from the disk center. Vials positioned at the periphery of the adapters (at a distance r_1) experiment most force, vials positioned more toward the center are at a distance r_2 and experience less force. This movement, referred to as *lunar*[24] translates into a net force different from the one typically experienced in a conventional planetary mill.

Several other types of mills are commercially available, depending on the targeted application, also other than chemical synthesis. Their description is out of the scope of this book, as for the pilot and manufacture facilities (some of them processing over 1000 kg), the focus being the description of mills commonly found in research laboratories and suitable for pedagogic purposes at undergraduate level,[25] as illustrated via selected examples described in Chapter 4.

However, worth mentioning is an approach to scalability based on continuous process by twin-screw extrusion (Figure 2.1, panel e). Mechanochemical tool is considered as the solid-state equivalent of solution-based flow reactors, it was successfully used for the preparation of MOFs[26] and APIs[27] at kg/h scale. Two screws (co- or counterrotating) transport the (solid) reagents across the barrel (*conveying zones*), while shear and compression forces are exerted in the *kneading zones*. The reagent-feed rate, the screw profile, the barrel length (and its temperature) and the residence time across the barrel are usually the parameters to adjust for optimizing the reaction.

In this regard, already for laboratory-scale processes, several independent studies describe how chemical, milling and processing parameters[28] influence the outcome of a mechanochemical reaction, as illustrated in Figure 2.3.[29] In addition, the

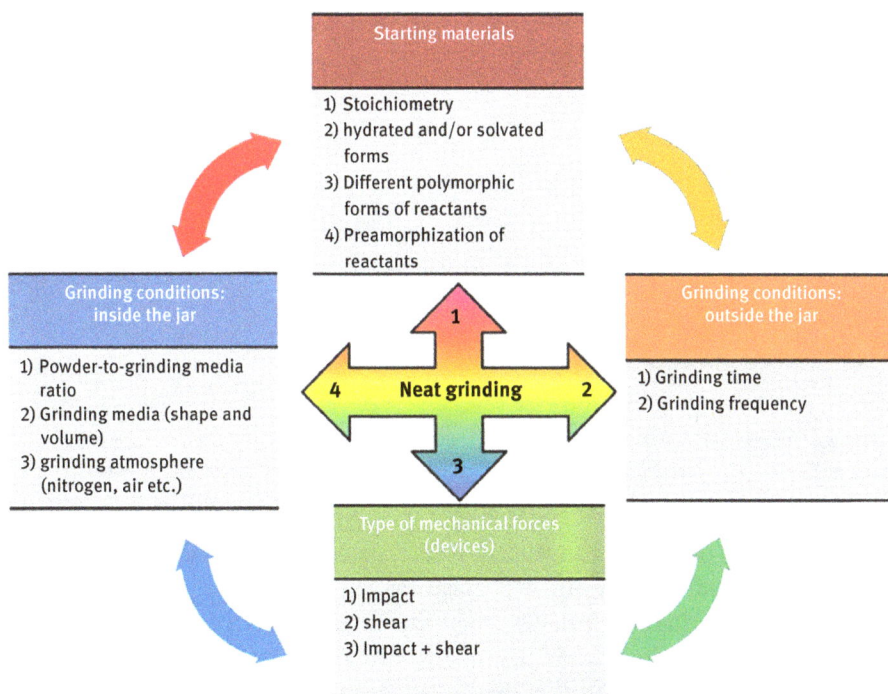

Figure 2.3: Some of the factors affecting the outcome of a neat mechanochemical reaction. Image reproduced from reference[29] with the permission of Elsevier.

energy inputs can be modulated also by modifying the number and the size of the balls, the frequency of collisions (stress frequency) or selecting milling materials with different hardness and density [polymethylmethacrylate (PMMA) 1.18 g cm^{-3}, polytetrafluoroethylene 2.2 g cm^{-3}, agate 2.64 g cm^{-3}, stainless steel 7.8 g cm^{-3}, zirconium oxide 5.68 g cm^{-3}, tungsten carbide 15.63 g cm^{-3}]. Generally speaking, denser materials deliver greater kinetic energy during the milling process. The selection of appropriate milling media also depends on the applications. The jar/ball material can promote a reaction,[30–31] but also become responsible of product contamination by metal leaching[32] or the rate of wearing of the specific material.[33]

In contrast to "dry milling" (also known as neat grinding), techniques making use of small amounts of liquid and/or solid additives have been introduced, presenting the advantages of increasing the reaction rate and driving the outcome of the reaction, independently on the solubility of the reactants in that solvent. For example, the preparation of the active pharmaceutical ingredient tolbutamide leads to two different polymorphs depending if grinding occurs in neat conditions or using a liquid-assisted grinding procedure (LAG) in the presence of a few microliters of nitromethane.[25]

LAG reactions are characterized empirically with the η(eta) parameter , expressed in µL/mg, and defined as the amount of added liquid (in µL) to the total weight of solid reactants (in mg).[29] The formal definition allows to distinguish quantitatively among neat grinding, LAG, slurry and reactions in solution (Figure 2.4).

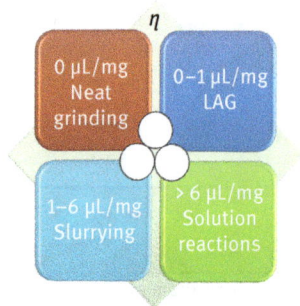

Figure 2.4: Ranges of η(eta)values. The reactions activated by ball milling are herein represented using the formalism of three circles grouped in a triangular arrangement (⚛), first introduced by Hanusa.[34]

The modification of the mechanochemical reaction environment occurs also by ion- and liquid-assisted grinding (ILAG),[35] used for the activation of metal oxides, or polymer-assisted grinding (POLAG)[36] and seeding-assisted grinding (SEAG),[37] allowing to control the formation of new pharmaceutical crystal forms.

The scope of mechanochemical transformations is now covering almost all areas of chemistry while advanced milling techniques employing additives have added new layers into control and selectivity of mechanochemical milling. Systematic use of these additives enhancing mechanochemical reactivity requires understanding of re-action mechanisms, to avoid that the reaction vessel is a black box where reactants are put in and products are collected. For the purpose of tracking the reaction course, milling can be periodically interrupted and the reaction vessel opened to take a sample of the reaction mixture, which will be analyzed *ex situ* using, for example, infrared spectroscopy, X-ray powder diffraction or some other analytical tech-nique (Figure 2.5). Ideally, time delay between sampling and analysis should be as short as possible to prevent possible transformations in the sample before the analysis. These transformations can for example, arise from the interaction of the sample with air and could be prevented by handling in a glove box under inert atmosphere. However, in some circumstances the glove box is not helpful. Some reactions (this can even be a majority of reactions of softer materials such as or-ganic materials) are not quenched when the mill is stopped, but the reactants, well mixed and finely ground, continue to react. In such a case, any reaction pro-file obtained by ex situ analysis will not be able to provide an accurate progress of the reaction in time.

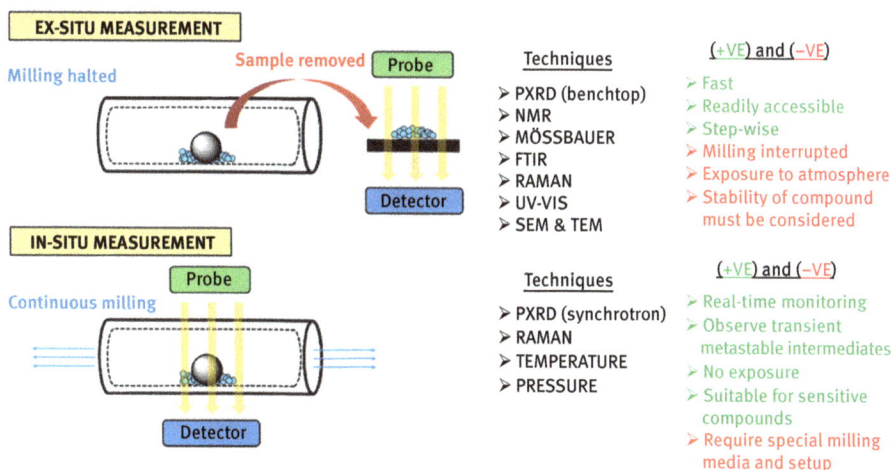

Figure 2.5: Schematic illustration of *in situ* or *ex situ* characterization techniques used in mechanochemical experiments. Image reproduced from reference[38] with permission of the Royal Society of Chemistry and Tecknoscience.[11]

Furthermore, sampling of the reaction mixture removes a part of the sample which may also alter mechanochemical reactivity by causing different rheological proper-ties of the milled sample – the sample is not milled in the same way when its amount decreases in the reaction vessel. Finally, milling is self-heating because the kinetic energy of the milling balls is transformed into internal energy of the system, during the collisions. Assuming that milling is started at room temperature, which is the case in the greatest majority of milling experiments, temperature of the milling assembly will rise. Stopping the milling process for sampling will also stop energy input and will lead to cooling. So, a mechanochemical reaction that is stopped and restarted will not have the same temperature profile as an uninterrupted reaction, which could easily affect reaction kinetics and mechanism. While this problem could in principle be remedied by repeating the same reaction, without sampling until the end of the reaction process, experience shows that mechanochemical reactions may not always be exactly reproducibile.[7]

We have now listed several problems encountered when one attempts to track mechanochemical reaction by using *ex situ* analysis. An *in situ* analysis on the other hand would ideally track the changing composition of the reaction mixture not only without opening of the reaction vessel, but also without interrupting the milling process. This is particularly challenging since the reaction vessel, being on a vibratory or a planetary ball mill, is moving rapidly. In addition, the reaction vessel is closed and is most often made from hard and nontransparent materials such as steel or a ceramic.

Nevertheless, two techniques for *in situ* reaction monitoring currently exist. One is based on X-ray diffraction,[39] while the other is based on Raman spectroscopy[40] (Figure 2.6). The two techniques work hand-in-hand, they are independent, and are

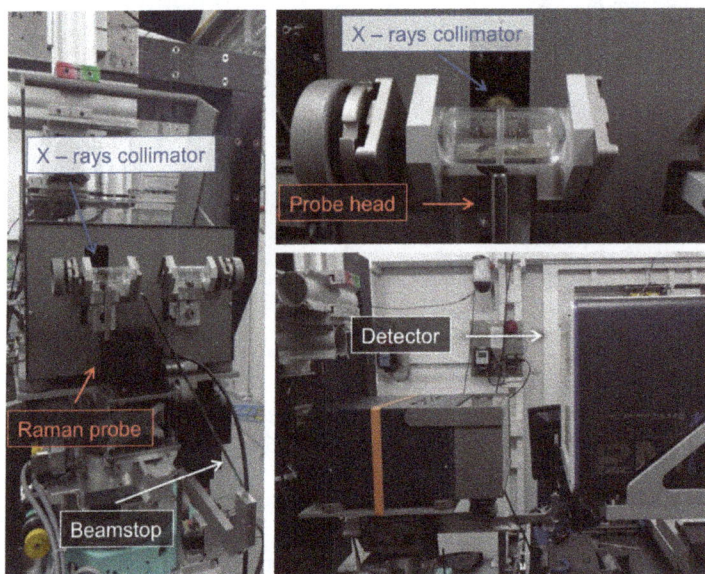

Figure 2.6: Setup for tandem *in situ* monitoring of mechanochemical milling reactions on a vibratory ball mill using X-ray diffraction and Raman scattering at the ESRF – The European Synchrotron beamline ID15A. Raman probe is approaching from below the reaction vessel and the X-ray beam is passed through the lower part of the inside chamber of the reaction vessel. Reproduced from the PhD. thesis of Stipe Lukin (University of Zagreb, 2019).

ideally used in tandem where the X-ray diffraction is performed simultaneously with Raman scattering.[41–42] For this purpose, the Raman laser is positioned to illuminate the same portion of the sample while the X-ray beam is passing through (Figure 2.6). In that way, the diffraction signal and the Raman scattering signal will come from the same portion of the sample which is important if the two are compared. If the milled sample is fully crystalline, the two reaction profiles derived from X-ray diffraction and Raman spectroscopy should exhibit a good match, but if, for example, the reaction mixture becomes partially amorphous, the reaction profiles from diffraction and Raman may differ (Figure 2.7).

This will be due to X-ray diffraction being sensitive only to crystalline species while Raman scattering collects signal from both the crystalline and amorphous parts of the sample. These two approaches have recently shed light on the fundamental understanding of mechanochemical mechanisms, broadening the potential of possible use of mechanochemistry in several other areas of application. *In situ* monitoring by Raman spectroscopy holds more promise to become available as a standard technique in mechanochemical laboratories because, though it may need a dedicated Raman spectrometer having a flexible Raman probe, it is still a laboratory technique, while X-ray diffraction requires access to a synchrotron X-ray source. Both techniques most often employ milling vessels made from polymethylmetacrylate (PMMA, Plexiglass, Perspex). A transparent reaction vessel is essential for Raman monitoring since the incident laser light must penetrate the reaction vessel wall to interact with the sample and the scattered radiation needs again to exit the vessel before being collected and transduced to the spectrometer. PMMA vessel is also beneficial for in situ X-ray diffraction as it is amorphous and does not give diffraction peaks but "just" increases the background.

Other than the two mentioned techniques to track the chemical composition of the milled reaction mixture, monitoring of temperature and pressure may also provide essential pieces of information to understanding of the milling process. As mentioned above, milling is self-heating and the temperature after initiation of milling will rise until heat dissipation to the surroundings equals dissipation of kinetic energy of the milling assembly. In such a case, temperature profiles for all mechanochemical reactions would look pretty much the same and would primarily depend on the amount of the material in the reaction vessel. However, properties of the material that is being milled are also important as different materials do not equally absorb kinetic energy of the milling balls. It thus happens that a steady-state temperature reached during milling of one material, after it transforms during milling, is changed to a new steady state. The change in temperature is most often too big to be a consequence of reaction enthalpy and is rather dominated by this change in the ability of a material to absorb kinetic energy of the milling assembly.[43]

Monitoring pressure during milling is particularly useful for solid-gas reactions where the gaseous reactant is charged in the reaction vessel and the progress of the

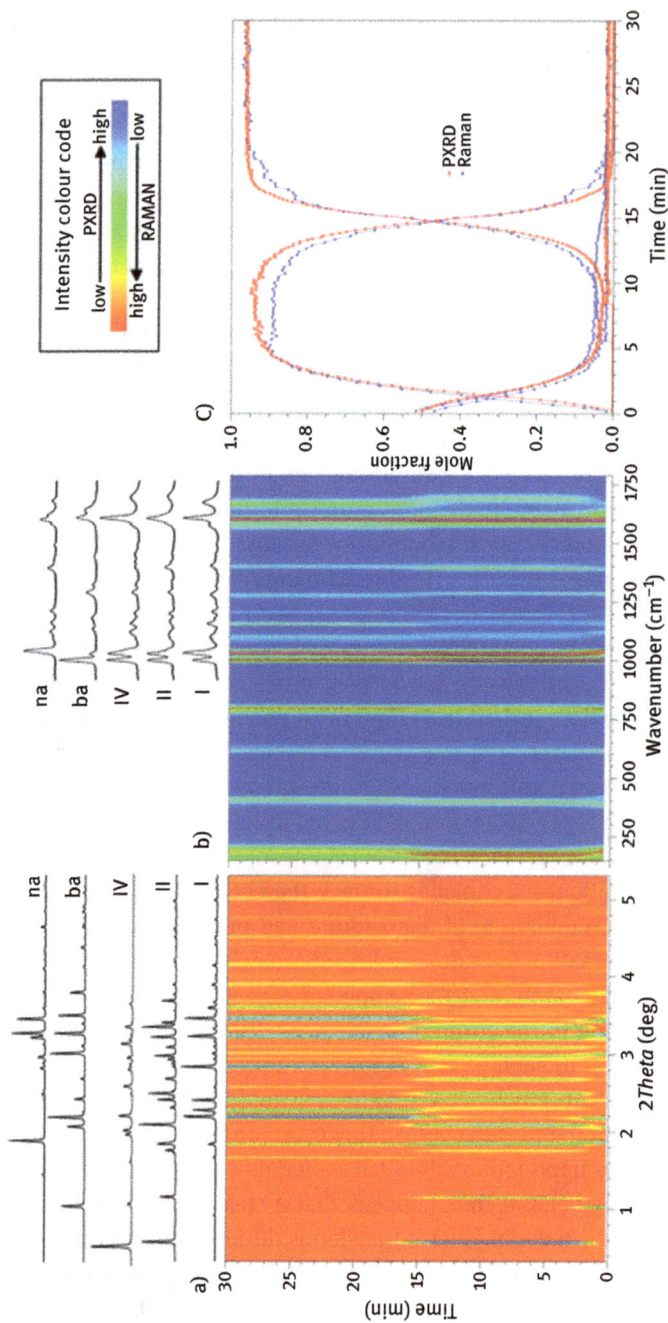

Figure 2.7: Two-dimensional plots of time-resolved (a) X-ray diffraction monitoring and (b) Raman spectroscopy monitoring for the formation of a cocrystal between benzoic acid and nicotinamide. In total, five species were identified – other than the two reactants, three polymorphs of their cocrystal appeared during the course of milling. (c) Diffraction- and Raman-derived reaction profiles show a good match due to the reaction mixture being fully crystalline. Reproduced from reference[41] with permission of Wiley.

reaction is monitored by a change in pressure. This was often applied to reactions involving gaseous hydrogen.

In situ monitoring techniques have thus far revealed a dynamic reaction environment with fast reactions, intermediates and reaction kinetics and mechanisms highly sensitive to slight changes in reaction conditions. Monitoring of the reaction course and observation of intermediates may allow for their efficient isolation as one can stop the milling at an appropriate time. It has thus been possible to identify, by crystal structure solution, several interesting intermediates including some elusive compounds which were previously presumed as intermediates, but were impossible to isolate from solution[44] or have been unexpectedly discovered.[7] That a mechanochemical environment can influence product selectivity is now well recognized and examples can be found in a review article.[9] However, what remains largely unknown is the underlying mechanism that causes these differences in reactivity. The mechanistic framework of mechanochemical milling reactions has barely been scratched so far and a huge body of work remains to be done before planning of milling experiments will not be based on experience.

Conclusion

Why not to contribute in advancing the sustainable thinking now? Instead of optimizing a reaction in solution, start exploring directly the process by mechanochemistry. If you do not have a mill at hand, maybe at your university (institute or company) there is one. Search around: ball mills and other types of mills find applications in forensic science, geology, biology, food science, pharmaceutical formulation and powdering materials. The scope of mechanochemical transformations is wide and it is quite possible that a reaction you are interested in has already been tested by mechanochemistry. The field is nevertheless young and a lot awaits to be discovered in terms of chemistry and applications, but also in developing instruments and methodology.

References

[1]   The term was introduced for the first time by Paracelsus in the sixteenth Century. See also: Joly, B. The alkahest, universal dissolvent or when theory turns an impossible practice into the imaginable. Rev. Hist. Sci. Paris 1996, 49, 305–344.
[2]   Ostwald, W. In the Fundamental Principles of Chemistry: an Introduction to All Text-Books of Chemistry. Longmans, Green and Co, New York, 1917.
[3]   James, S. L., Adams, C. J., Bolm, C., Braga, D., Collier, P., Friščić, T., Grepioni, F., Harris, K. D. M., Hyett, G., Jones, W., Krebs, A., Mack, J., Maini, L., Orpen, A. G., Parkin, I. P.,

Shearouse, W. C., Steed, J. W., Waddell, D. C. Mechanochemistry: opportunities for new and cleaner synthesis. Chem. Soc. Rev. 2012, 41, 413–447.

[4] Do, J.-L., Friščić, T. Chemistry 2.0: developing a new, solvent-free system of chemical synthesis based on mechanochemistry. Synlett 2017, 28, 2066–2092.

[5] Tan, D., Loots, L., Friščić, T. Towards medicinal mechanochemistry: evolution of milling from pharmaceutical solid form screening to the synthesis of active pharmaceutical ingredients (APIs). Chem. Commun. 2016, 52, 7760–7781.

[6] Shi, Y. X., Xu, K., Clegg, J. K., Ganguly, R., Hirao, H., Friščić, T., Garcia, F. The first synthesis of the sterically encumbered adamantoid phosphazane $P_4(N(t)Bu)_6$: enabled by mechanochemistry. Angew. Chem. Int. Ed. 2016, 55, 12736–12740.

[7] Katsenis, A. D., Puškarić, A., Štrukil, V., Mottillo, C., Julien, P. A., Užarević, K., Pham, M.-H., Do, T.-O., Kimber, S. A. J., Lazić, P., Magdysyuk, O., Dinnebier, R. E., Halasz, I., Friščić, T. In situ X-ray diffraction monitoring of a mechanochemical reaction reveals a unique topology metal-organic framework. Nat. Commun. 2015, 6, 6662–6669.

[8] Howard, J. L., Cao, Q., Browne, D. L. Mechanochemistry as an emerging tool for molecular synthesis: what can it offer?. Chem. Sci. 2018, 9, 3080–3094, and references cited therein.

[9] Hernandez, J., Bolm, C. Altering Product Selectivity by Mechanochemistry. J. Org. Chem. 2017, 82, 4007–4019.

[10] Hernández, J. G., Halasz, I., Crawford, D. E., Krupička, M., Baláž, M., André, V., Vella-Zarb, L., Niidu, A., García, F., Maini, L., Colacino, E. European research in focus: mechanochemistry for sustainable industry (MechSustInd). Eur. J. Org. Chem. 2020, 8–9.

[11] Baláž, M., Vella-Zarb, L., Hernandez, J., Halasz, I., Crawford, D. E., Krupička, M., André, V., Niidu, A., Garcia, F., Maini, L., Colacino, E. Mechanochemistry: a disruptive innovation for the industry of the future. Chem. Today 2019, 37, 32–34.

[12] The importance of mechanochemistry has been recently acknowledged by the International Union of Pure and Applied Chemistry (IUPAC) and by the European Cooperation in Science and Technology (COST) funding COST Action: CA18112 – Mechanochemistry for Sustainable Industry (MechSustInd, http://www.mechsustind.eu/). See also: Gomollón-Bel, F. Ten Chemical Innovations That Will Change Our World. Chem. Int. 2019, 41, 12–17.

[13] Ball Milling Towards Green Synthesis: Applications, Projects, Challenges. Stolle, A., Ranu, B., Ed. RSC Green Chemistry Series, 2015.

[14] Mechanochemistry: From Functional Solids to Single Molecule, Faraday Discuss. RSC, Cambridge, UK, 2014, Vol. 170.

[15] Margetic, D., Strukil, V. Mechanochemical Organic Synthesis. Elsevier, Amsterdam, NL, 2016, 386.

[16] Charnay, C., Porcheddu, A., Delogu, F., Colacino, E. New and up-and-coming perspectives for an unconventional chemistry: from molecular synthesis to hybrid materials by mechanochemistry. In: Green Synthetic Processes and Procedures, Edited by Ballini, R., Ed. RSC Green Chemistry Series (2019), Ch. 9.

[17] Porcheddu, A., Charnay, C., Delogu, F., Colacino, E. From solution-based non-conventional activation methods to mechanochemical procedures: the hydantoin case. In: Non Traditional Activation Methods in Green and Sustainable Applications, Edited by Torok, B., Christian Schafer in 'Advances in Green and Sustainable Chemistry series' (series Editors Bela Torok and Timothy Dransfield), Ed. Elsevier (2020) – in print.

[18] Baláž, P., Achimovičová, M., Baláž, M., Billik, P., Cherkezova-Zheleva, Z., Criado, J. M., Delogu, F., Dutková, E., Gaffet, E., Gotor Martinéz, F. J., Kumar, R., Mitov, I., Rojac, T., Senna, M., Streletskii, A., Wieczorek-Ciurowa, K. Hallmarks of mechanochemistry: from nanoparticles to technology. Chem. Soc. Rev. 2012, 42, 7571–7637.

[19]   Levitas, V. I. Continuum mechanical fundamentals of mechanochemistry. In: High Pressure Surface Science and Engineering, Gogotsi, Y., Domnich, V., Eds. Bristol, Institute of Physics, 2004, 159–292.

[20]   Klika, V., Marsik, F. Coupling effect between mechanical loading and chemical reactions. J. Phys. Chem. B 2009, 113, 14689–14697.

[21]   Leonardi, M., Villacampa, M., Menéndez, J. C. Multicomponent mechanochemical synthesis. Chem. Sci. 2018, 9, 2042–2064.

[22]   Chen, L., Leslie, D., Coleman, M. G., Mack, J. Recyclable heterogeneous metal foil-catalyzed cyclopropenation of alkynes and diazoacetates under solvent-free mechanochemical reaction conditions. Chem. Sci. 2018, 9, 4650–4661.

[23]   Martina, K., Rotolo, L., Porcheddu, A., Delogu, F., Bysouth, S. R., Cravotto, G., Colacino, E. High throughput mechanochemistry: application to parallel synthesis of benzoxazines. Chem. Commun. 2018, 54, 551–554.

[24]   Bysouth, S. R. US2006/0175443 A1, 2006.

[25]   Colacino, E., Dayaker, G., Morère, A., Friščić, T. Introducing students to mechanochemistry via environmentally friendly organic synthesis using a solvent-free mechanochemical preparation of the antidiabetic drug tolbutamide. J. Chem. Educ. 2019, 96, 766–771.

[26]   Crawford, D., Casaban, J., Haydon, R., Giri, N., McNally, T., James, S. L. Synthesis by extrusion: continuous, large-scale preparation of MOFs using little or no solvent. Chem. Sci. 2015, 6, 1645–1649.

[27]   Crawford, D.E, Porcheddu, A., McCalmont, A.S., Delogu, F., James, S.L., Colacino, E. Solvent-free, Continuous Synthesis of Hydrazone-Based Active Pharmaceutical Ingredients by Twin-Screw Extrusion. ACS Sustainable Chem. Eng. 2020, 8, 12230–12238.

[28]   Burmeister, C. F., Stolle, A., Schmidt, R., Jacob, K., Breitung-Faes, S., Kwade, A. Experimental and computational investigation of Knoevenagel condensation in planetary ball mill. Chem. Eng. Technol. 2014, 37, 857–864.

[29]   Hasa, D., Jones, W. Screening for new pharmaceutical solid forms using mechanochemistry: a practical guide. Adv. Drug. Deliv. Rev. 2017, 117, 147–161.

[30]   Fulmer, D. A., Shearouse, W. C., Medonza, S. T., Mack, J. Solvent-free Sonogashira coupling reaction via high speed ball milling. Green Chem. 2009, 11, 1821–1825.

[31]   Sawama, Y., Kawajiri, T., Niikawa, M., Goto, R., Yabe, Y., Takahashi, T., Marumoto, T., Itoh, M., Kimura, Y., Monguchi, Y., Kondo, S.-I., Sajiki, H. Stainless-steel ball-milling method for hydro-/deutero-genation using H_2O/D_2O as a hydrogen/deuterium source. ChemSusChem 2015, 8, 3773–3776.

[32]   Štefanić, G., Krehula, S., Štefanić, I. The high impact of a milling atmosphere on steel contamination. Chem. Commun. 2013, 49, 9245–9247.

[33]   Rak, M. J., Saadé, N. K., Friščić, T., Moores, A. Green Chem. 2014, 16, 86–89.

[34]   Rightmire, N. R., Hanusa, T. P. Advances in organometallic synthesis with mechanochemical methods Dalton Trans. 2016, 45, 2352–2362.

[35]   Friščić, T., Reid, D. G., Halasz, I., Stein, R. S., Dinnebier, R. E., Duer, M. J. Ion- and Liquid-Assisted Grinding: improved mechanochemical synthesis of metal–organic frameworks reveals salt inclusion and anion templating. Angew. Chem. Int. Ed. 2010, 49, 712–715.

[36]   Hasa, D., Schneider Rauber, G., Voinovich, D., Jones, W. Cocrystal formation through mechanochemistry: from neat and liquid-assisted grinding to polymer-assisted grinding. Angew. Chem. Int. Ed. 2015, 54, 7371–7375.

[37]   Cinčić, D., Brekalo, I., Kaitner, B. Solvent-free polymorphism control in a covalent mechanochemical reaction. Cryst. Growth Des. 2012, 12, 44–48.

[38]   Tan, D., Garcia, F. Main group mechanochemistry: from curiosity to established protocols. Chem. Soc. Rev. 2019, 48, 2267–2496.

[39] Friščić, T., Halasz, I., Beldon, P. J., Belenguer, A. M., Adams, F., Kimber, S. A. J., Honkimäki, V., Dinnebier, R. E. Real-time and in situ monitoring of mechanochemical milling reactions. Nat. Chem. 2013, 5, 66–73.

[40] Gracin, D., Štrukil, V., Friščić, T., Halasz, I., Užarević, K. Laboratory real-time and in situ monitoring of mechanochemical milling reactions by Raman spectroscopy. Angew. Chem., Int. Ed. 2014, 53, 6193–6197.

[41] Lukin, S., Stolar, T., Tireli, M., Blanco, M. V., Babić, D., Friščić, T., Užarević, K., Halasz, I. Tandem in situ monitoring for quantitative assessment of mechanochemical reactions involving structurally unknown phases. Chem. Eur. J. 2017, 23, 13941–13949.

[42] Batzdorf, L., Fischer, F., Wilke, M., Wenzel, K., Emmerling, F. Direct In Situ Investigation of Milling Reactions Using Combined X-ray Diffraction and Raman Spectroscopy. Angew. Chem. Int. Ed. 2015, 54, 1799–1802.

[43] Užarević, K., Ferdelji, N., Mrla, T., Julien, P. A., Halasz, B., Friščić, T., Halasz, I. Enthalpy vs. friction: heat flow modelling of unexpected temperature profiles in mechanochemistry of metal–organic frameworks. Chem. Sci. 2018, 9, 2525–2532.

[44] Strukil, V., Gracin, D., Magdysyuk, O. V., Dinnebier, R. E., Friščić, T. Trapping reactive intermediates by mechanochemistry: elusive aryl N-thiocarbamoylbenzotriazoles as bench-stable reagents. Angew. Chem. Int. Ed. 54, 8440–8442.

Alessandra Scano, Guido Ennas

3 Rediscovering mechanochemistry for inorganic materials

This chapter deals with the use of mechanochemistry in the synthesis of inorganic materials. An assortment of routes for the mechanosynthesis of elementary substances, alloys, oxides, sulfides and nanocomposites at the laboratory scale is presented.

As mentioned in Chapter 1, mechanochemistry of inorganic materials represents the deepest routed area of mechanochemical synthesis, starting in 1820 with the work of Faraday who described reduction of silver chloride by the "dry way" of inducing reactions.[1]

Modern mechanochemistry of inorganic materials includes (1) transformation from crystalline to disordered amorphous phases, (2) preparation of alloys via mechanical alloying or mechanical milling (MM), (3) mechanosynthesis of oxides, halides, sulfides, nitrides and (4) preparation of composites.[2]

Mechanosynthesis related to the preparation of the abovementioned types of materials, with some specific examples, will be described in this chapter. First, milling of an elementary substance and the condition to transform a crystalline to an amorphous phase are reported. The latter will be also discussed in relation to the preparation of alloys.

To verify the success of the proposed synthesis, the powder X-ray diffraction technique is suggested to characterize the samples crystallographically while electron microscopy is well suited to provide information on the sample morphology.

3.1 Elemental substance

MM of an elemental substance leads to the decrease of grain sizes, also referred to as refinement process, until a nanostructure is obtained. The capability of the crystalline phase to sustain a nanostructured state during the mechanical process is the key point in the amorphous formation. In this regard, literature reports detailed studies about the MM of various elementary substances such as Si and Ge, highlighting that the transformation of a crystalline phase into an amorphous takes place when a critical downsizing of grains is reached in the milling process. Below this critical size, the surface energy becomes higher than the volume energy and amorphization takes place.[3]

Elemental substances can be obtained by mechanochemical synthesis using multiple reactants and products, but with different solubilities. An example is the production of silver nanoparticles (NPs) by reduction of AgCl using either Na or Cu,

https://doi.org/10.1515/9783110608335-003

with the CuCl and NaCl byproducts removed by washing with aqueous ammonia.[4, 5] In this case, stoichiometric equivalent molar ratio of analytical grade AgCl and metallic Cu (<325 mesh) are milled in a planetary ball mill (Table 3.1). Ascorbic acid, being a reducing agent, is used as an additive to help the reaction between the starting materials (AgCl and Cu):

Table 3.1: Process parameters for the synthesis of Ag NPs by mechanochemistry.

Parameters of the milling process	Method
Type of mill	Planetary mill
Jar volume (mL)	45
Jar material	ZrO_2
Number of balls	7
Balls diameter (mm)	15
Weight of each ball (g)	n.r. [a] (10)[b]
Milling speed	700 rpm[d]
Milling time (h)	10
Milling and resting period (min)	n.r. [a]
Grinding additive	Ascorbic acid
Reaction atmosphere	n.r.[a] (inert)[c]
Reaction scale (mmol)	10

[a]n.r., not reported;
[b]calculated weight (using $d_{ZrO_2} = 5.7 \text{ g/cm}^3$);
[c]inert, better in nitrogen or argon gas;
[d]rpm, revolutions per minute.

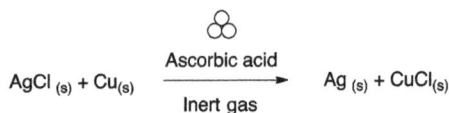

$$AgCl_{(s)} + Cu_{(s)} \xrightarrow[\text{Inert gas}]{\text{Ascorbic acid}} Ag_{(s)} + CuCl_{(s)}$$

Equation 3.1. Ag NPs by mechanochemistry. Reaction conditions for method: AgCl 10 mmol, Cu 10 mmol, ascorbic acid 2.0 g (11.4 mmol).

After milling, a leaching treatment of the ground samples is necessary in order to remove the secondary products, CuCl and the residue of unreacted AgCl. Under a chemical fume hood, 1 g of the milled sample is dispersed in 100 mL of 1.0 M aqueous ammonia in a 250 mL conical flask, and the slurry is stirred by a magnetic bar. The following reactions occur:

$$CuCl_{(s)} + 2 NH_{3 (aq)} \longrightarrow [Cu(NH_3)_2] Cl_{(aq)}$$

$$AgCl_{(s)} + 2 NH_{3 (aq)} \longrightarrow [Ag(NH_3)_2] Cl_{(aq)}$$

After leaching, slurry is filtered by using a membrane filter (cellulose acetate, pore size 0.2 µm) and washed with small amounts of ammonia water solution (3×5 mL) and distilled water (3×5 mL).

3.2 Alloys

Homogeneous alloys are intermetallic phases or solid solutions.[6] They can be prepared by solid-state reactions (SSR) by using different routes: (1) cold working of intercalated foils, (2) mechanical mixing of pure elements in their powdered form (mechanical alloying, MA), (3) grinding of intermetallic compounds (MM)[7, 8] and (4) grinding of already prepared alloys and elements (MA).[9–11] During the MA and MM processes in high-velocity ball mills, the energy transfer from the milling tools to the milled powder is achieved. First, the action of repetitive milling collisions induces a significant reduction in crystallite and particle sizes of the milled precursors (comminution step) that, in some case, can amorphize by further prolonged milling.[12–14] Preparation of alloys by MA or MM generally require long milling times (from 24 up to even 300 h).[9]

When MA of mixed pure elementary substances is carried out, some steps of the SSR can be identified:
i)   the induction period, consisting in the development of stable nuclei;
ii) acceleratory period during which the growth of such nuclei is achieved, sometimes accompanied by further nucleation, until the maximum solubility rate of one element into the second element;
iii) decay period, which represents the end of the nuclei expansion, due to impingement and consumption of reactants;
iv) deceleration period, which continues until reaction completion.[15]

Literature reports on the dependence of the final products of the milling process from the morphology of the starting powders and from the main milling parameters. Concerning the starting powders and referring to ductile materials, the formation of lamellas and the modification of the lamellar spacing is influenced by the particle morphology and size of the powders. In fact, starting from powders with small particle size and narrow size distribution the processing time decreases.[16]

Regarding the milling parameters, it has been found that the transformed mass fraction depends on the total mechanical work done on the processed powders. In particular, the conversion degree and rates are strongly related to the impact energy, collision frequency (correlated to the mean free path of the balls) and powder charge. Moreover, the use of different mills or different conditions of the same mill results in markedly different reactions pathways.[17–19]

General routes to prepare alloys are shown in Scheme 3.1.

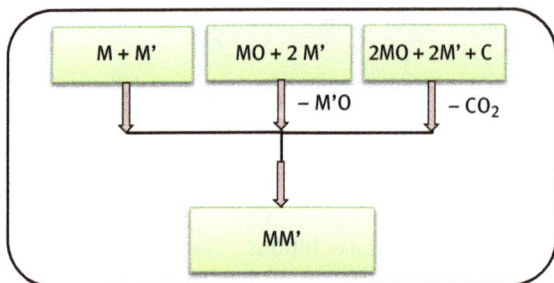

| M + M' | MO + 2 M' | 2MO + 2M' + C |

– M'O   – CO_2

MM'

Scheme 3.1: General ball milling routes to prepare alloys. M represents metal (*e.g.*, Fe, Zr, Co, Cu, Mo, Mn, Al, Ni and Ti), MO represents metal oxide (*e.g.*, CuO, ZnO and PbO).

The simplest way to prepare binary alloys consists of the combination of two pure metals (Scheme 3.1). Some examples were reported in the literature where the preparation of Fe–Zr, Co–Cu, Fe–Mo, Mn–Al and Ni–Ti alloys has been described.[9, 10, 20–24] In some cases, amorphous phases instead of solid solutions were obtained due to favorable thermodynamic and kinetic conditions. Generally, reactions of metals require long grinding times.

Combinations of a metal and a semimetal have been used for boron or silicon-containing binary alloys in the Co–B, Ni–B and Pd–Si systems[25–27] and ternary alloys in the Ni–Nb–B and Ti–Al–B systems.[28, 29] As an example, the mechanochemical reaction between metallic nickel and boron powders to obtain Ni_2B in a planetary ball mill apparatus is reported (Table 3.2). After a few milling hours, the SSR reaches a steady state where the orthorhombic nanocrystalline phase o-Ni_3B and an amorphous phase are observed. After 55 h milling time, the formation of the tetragonal intermetallic compounds t-Ni_2B, with a rather narrow particle size distribution and an average dimension of about 10 nm, is almost complete.

Ternary alloys have also been prepared starting from three different metals.[30] When pure metals are used as starting materials, the reaction needs to be carried out under inert atmosphere (typically argon or nitrogen gas) in order to avoid oxidation of the metal and the produced alloy.

Table 3.2: Process parameters for mechanochemical reaction to synthesize Ni–B alloy.

Key parameters of the milling process	Method
Type of mill	Planetary mill
Jar volume (mL)	250
Jar material	SS [a]
Number of balls	100
Balls diameter (mm)	8
Weight of each ball (g)	2.09
Milling speed	250 rpm
Milling time (h)	55
Milling and resting period (min)	5 and 5
Grinding additive	–
Reaction atmosphere	Ar
Reaction scale (mmol)	156

[a]SS, stainless steel

MA in combination with single exchange redox reactions represents an alternative route to obtain alloys. One of the metal reagents (M) is substituted by a binary oxide (MO), which is reduced by the second metal (M') or by a third reducing component (typically carbon). This alternate way is very suitable for the production of alloys that are usually difficult to be obtained by conventional methods. Mg–Ti alloy is a typical example. In fact, conventional casting methods cannot achieve complete mixing of Ti and Mg due to the boiling point of Mg (1090 °C), which lies well below the melting point of Ti (1668 °C). In this case, Mg–Ti alloy was obtained by MA of a binary oxide powder (TiO_2) and a reducing agent (Mg), eliminating the secondary product MgO and the unreacted reagents by using sequential hydrometallurgical means.[31]

When carbon is introduced as a third component in the starting mixture, the problem to eliminate secondary products is overcome because of easy removal of gaseous CO_2. An example is the synthesis of brass from CuO, ZnO and PbO in the presence of graphite.[32]

3.3 Oxides

Mechanochemical synthesis of oxides has represented an important area of solid-state chemistry because it offers the possibility to prepare oxides by different pathways,[2, 3] as illustrated in Scheme 3.2:

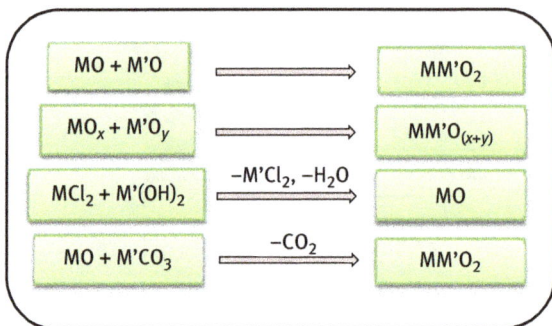

Scheme 3.2: General ball-milling routes to prepare metal oxides. M and M' = metals (*e.g.*, Fe, Ni, Ti and Zn).

Mechanochemistry of two binary oxides is the simplest way to obtain nanosized mixed oxides such as spinels and perovskites.[2, 33] In some cases, a thermal treatment of the milled products is necessary to get the desired oxide. For example, the synthesis of ultrafine magnetic particles of nickel ferrite ($NiFe_2O_4$) via BM is reported. Two starting powder materials, nickel oxide NiO and hematite α-Fe_2O_3, are mixed with a molar ratio of 1:1 according to the following equation reaction:

$$NiO_{(s)} + \alpha\text{-}Fe_2O_{3\,(s)} \xrightarrow[\text{Air}]{} NiFe_2O_{4\,(s)}$$

Equation 3.2. Nickel ferrite ($NiFe_2O_4$) via mechanochemistry. Reaction conditions for method: NiO (43 mmol), α-Fe_2O_3 (43 mmol).

After 8 h milling of the mixture, only $NiFe_2O_4$ phase is detected, indicating that the reaction is complete. Thermal treatment of the as-milled powder at 700 °C for 1 h improves the crystallization of $NiFe_2O_4$ NPs. The final average crystal size is about 23 nm.

Starting point to prepare spinels and perovskites could also be an elemental metal in combination with a metal oxide. This reaction is thermodynamically and kinetically favored and requires to be carried out in inert atmosphere. Some examples are (1) preparation of $FeTiO_3$ and $FeTiO_4$ using titanium as the reducing metal,[34] (2) the formation of Fe_2GeO_4 using iron,[35] (3) the synthesis of $ZnFe_2O_4$ starting from zinc[36] and of (4) $FeAl_2O_4$ from aluminum.[37] The single-step synthesis of Fe_2GeO_4 via high-energy BM of the mixture of three precursors is reported as an example (eq. (3.3)):

$$2 \; \alpha\text{-Fe}_2\text{O}_{3\,(s)} + 2 \; \text{Fe}_{(s)} + 3 \; \text{GeO}_{2\,(s)} \xrightarrow[\text{Argon gas}]{} 3 \; \text{Fe}_2\text{GeO}_{4\,(s)}$$

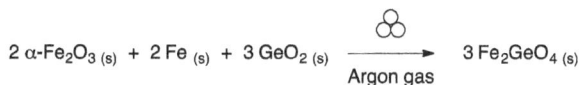

Equation 3.3. Preparation of iron germanate Fe_2GeO_4 via mechanochemistry. Reaction conditions for method: $\alpha\text{-Fe}_2O_3$ (26.85 mmol), GeO_2 (40.3 mmol). Powders of the three $\alpha\text{-Fe}_2O_3$, Fe and GeO_2 reactants, in the molar ratio of 2:2:3, are subjected to uninterrupted milling for up to 2 h in a planetary ball mill at room temperature and under an argon atmosphere. Milling conditions are given in Table 3.3.

Table 3.3: Process parameters for mechanochemical reaction to synthesize Fe_2GeO_4.

Parameters of the milling process	Method
Type of mill	Planetary mill
Jar volume (mL)	250
Jar material	WC[a]
Number of balls	24
Balls diameter (mm)	10
Weight of each ball (g)	2.97
Milling speed	600 rpm
Milling time (h)	2
Milling and resting period (min)	uninterrupted
Grinding additive	NaCl
Reaction atmosphere	Ar
Reaction scale (mmol)	40.0

[a]WC, tungsten carbide.

An alternative way consists in the reaction between metal salt (*e.g.*, chlorides) and metal hydroxides (alkaline or alkaline earth metals). The reaction gives rise to several products with different solubilities that could be the basis for their separation.[38] An example is the reaction to obtain manganese ferrite NPs (eq. (3.4)).

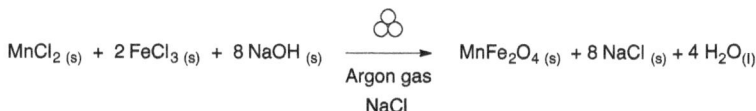

$$\text{MnCl}_{2\,(s)} + 2 \; \text{FeCl}_{3\,(s)} + 8 \; \text{NaOH}_{(s)} \xrightarrow[\substack{\text{Argon gas} \\ \text{NaCl}}]{} \text{MnFe}_2\text{O}_{4\,(s)} + 8 \; \text{NaCl}_{(s)} + 4 \; \text{H}_2\text{O}_{(l)}$$

Equation 3.4. Preparation of manganese ferrite $MnFe_2O_4$ via mechanochemistry. Reaction conditions for method: $MnCl_2\cdot 4H_2O$ (5 mmol), $FeCl_3 \cdot 6H_2O$ (10 mmol), NaOH (40 mmol).

The NaCl byproduct can be removed by washing with water. Pulverized sodium hydroxide has to be previously prepared by milling for 5 min the commercial pellet material in the same shaker apparatus. Teflon-coated milling tools are used to prevent corrosion. BM conditions are given in Table 3.4. Sodium chloride is added to the mixture to limit crystal growth and to obtain small particles of an average size of 7.5 nm.

Table 3.4: Process parameters for mechanochemical synthesis of $MnFe_2O_4$.

Parameters of the milling process	Method
Type of mill	Mixer mill
Jar volume (mL)	32
Jar material	Teflon coated SS[a]
Number of balls	5
Balls diameter (mm)	10
Weight of each ball (g)	2.97 (ZrO_2)
Milling speed	1060 cpm[b]
Milling time (h)	1
Milling and resting period (min)	5 and 5
Grinding additive	NaCl (5.3 g)
Reaction atmosphere	Ar
Reaction scale (mmol)	5.0

[a]SS, stainless steel,
[b]cpm, cycles per minute.

Similar displacement reactions with salt products have been used to synthesize oxides such as ZnO, ZrO_2, Cr_2O_3, $LaCoO_3$ and Nb_2O_5.[2] The same pathway, generating MO or M_2O oxides and consisting in the mixing of alkaline and alkaline earth carbonates, can be used to obtain other binary compounds such as $LiFe_5O_5$, $CaTiO_3$ and $NaNbO_3$ or ternary oxides like $Ba_{1-x}Sr_xTiO_3$, where the secondary product is eliminated as CO_2.[2] The carbonate method has been used to synthesize more complex oxides (*e.g.*, $Ba_2ANb_5O_{15}$ where A = K, Na and Li). As example, the synthesis of $NaNbO_3$ is reported.[39] It can be carried out by milling a mixture of powdered Na_2CO_3 and Nb_2O_5, weighed to satisfy the molar ratio of Na:Nb of 1:1 (eq. (3.5)).

$$Nb_2O_{5\ (s)} + Na_2CO_{3\ (s)} \xrightarrow[\text{Air}]{\ \ 8\ \ } 2NaNbO_{3\ (s)} + CO_{2\ (g)}$$

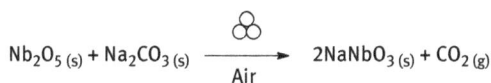

Equation 3.5. Sodium niobate $NaNbO_3$ via mechanochemistry. Reaction conditions for method: Nb_2O_5 (8 mmol), Na_2CO_3 (8 mmol).

Due to its hygroscopic nature, the Na_2CO_3 powder needs to be dried before use. The powder mixture is first homogenized in acetone using a planetary mill (200 rpm, 2 h), dried at 150 °C for 1 h and then subjected to BM. Two experiments adopting different milling conditions are reported in Table 3.5. It is suggested to use the faster method B.

Table 3.5: Process parameters for mechanosynthesis of $NaNbO_3$.

Parameters of the milling process	Method A	Method B
Type of mill	Planetary mill	Planetary mill
Jar volume (mL)	250	125
Jar material	YSZ[a]	WC[b]
Number of balls	40	16
Balls diameter (mm)	10	15
Weight of each ball (g)	7.91	26.7
Milling speed	200 rpm	300 rpm
Milling time (h)	400	96
Milling and resting period (min)	n.r. [c]	n.r. [c]
Grinding additive	–	–
Reaction atmosphere	Air	Air
Reaction scale (mmol)	80.7	13.4

[a]YSZ, yttrium-stabilized zirconia;
[b]WC, tungsten carbide;
[c]n.r., not reported.

3.4 Sulfides and other metal–nonmetal compounds

Metal–nonmetal compounds such as sulfides, selenides, halides and nitrides are very interesting materials because of their physicochemical properties and consequently their technological applications. Mechanochemistry is a suitable technique to prepare these materials starting by the direct combination of elements otherwise by reaction of their compounds, according to Schemes 3.3–Schemes 3.5.

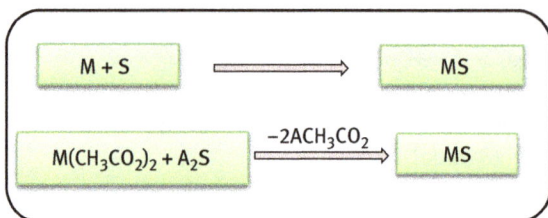

Scheme 3.3: General ball milling routes to prepare sulfides. M = metal (*e.g.*, Fe, Zn, Cd and Pb), A = alkali metal (*e.g.*, Na).

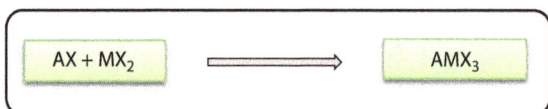

Scheme 3.4: General ball milling routes to prepare halides. M = metal (*e.g.*, Mg, Zn, Mn, Ni, Cu, Co and Fe), A = alkali metal (*e.g.*, Na and K).

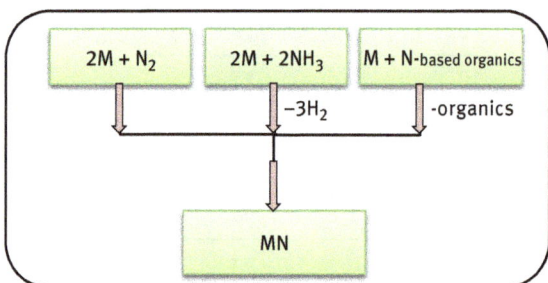

Scheme 3.5: General ball milling routes to prepare nitrides. M = metal (*e.g.*, Ti, Zr, V, Nb, Mo and Ga) or semimetal (*e.g.*, B and Si), N-based organics = urea, pyrazole and phenylenediamine.

In particular, sulfides exhibit a great variety of physicochemical properties due to structural defects related to cation vacancies, interstitial cations or anionic defects much more varied than in their oxide counterpart.

A wide range of sulfidic nanocrystals has been prepared by mechanosynthesis as reported in the literature. An interesting example is the quite rapid mechanosynthesis of Sb_2S_3 and Bi_2S_3 NPs starting from the corresponding metals and sulfur in a planetary ball mill.[3] An alternative strategy is the mechanosynthesis of sulfide NPs by SSRs of metal acetates (metal = Zn, Cd, Pb, etc.) with alkaline sulfides,[40] according to the following general reaction:

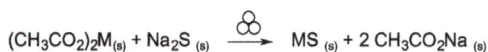

$$(CH_3CO_2)_2M_{(s)} + Na_2S_{(s)} \xrightarrow{} MS_{(s)} + 2\,CH_3CO_2Na_{(s)}$$

where at the end of the BM process, the solid sulfide metal NPs can be directly obtained by washing the unreacted precursors and the soluble product. In particular, the synthesis of ZnS NPs is achieved according to eq. (3.6):

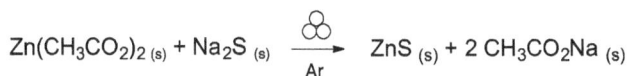

$$Zn(CH_3CO_2)_{2\ (s)} + Na_2S_{\ (s)} \xrightarrow[\text{Ar}]{} ZnS_{\ (s)} + 2\ CH_3CO_2Na_{\ (s)}$$

Equation 3.6. Zinc sulfide (ZnS) NPs via mechanochemistry. Reaction conditions for method: $Zn(CH_3CO_2)_2 \cdot 2\ H_2O$ (31 mmol), $Na_2S \cdot 9\ H_2O$ (31 mmol).

Table 3.6: Process parameters for mechanosynthesis of zinc sulfide (ZnS).

Parameters of the milling process	Method
Type of mill	Planetary mill
Jar volume (mL)	n.r. [a] (250 mL)[b]
Jar material	WC
Number of balls	50
Balls diameter (mm)	10
Weight of each ball (g)	n.r. [a] (8.2)[c]
Milling speed	500 rpm
Milling time (min)	10
Milling and resting period (min)	n.r. [a]
Grinding additive	–
Reaction atmosphere	Ar
Reaction scale (mmol)	31

[a]n.r., not reported;
[b]recommended volume;
[c]calculated value (using $d_{WC} = 15.6$ g cm^{-3}).

The purification of the as-milled powder is achieved by washing in water, decantation and drying (50 °C, 120 min).

3.5 Composites

A composite is a material made from two or more constituents that, when combined, produces a new material with chemical and physical properties different from the individual components. It differs from solid solutions because the individual components remain separate within the product material.

In the last decades, particular interest has been addressed to nanocomposite materials where a nanophase is dispersed in a matrix, which generally can be metallic, ceramic and more recently polymeric. Both nanophase and matrix are generally simultaneously synthesized during the process.[2]

Most of the studies related to the preparation of nanocomposites interested the embedding of oxides NPs in metal or alloy matrices to confer improved mechanical properties.

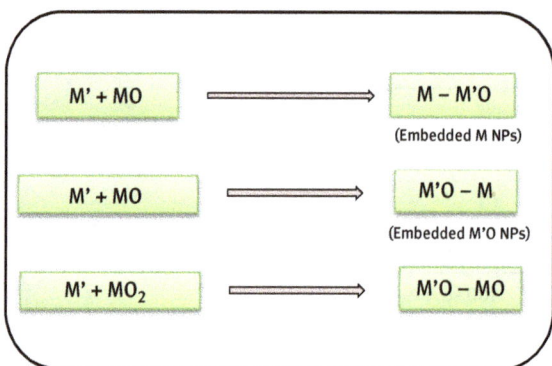

Scheme 3.6: General ball milling routes to prepare composites. M, metal (*e.g.*, Zn, Nb, Cu and Al), nonmetal (*e.g.*, C) or semimetal (*e.g.*, Si).

Typical examples include the preparation of oxide dispersion-strengthened alloys by BM for gas turbine applications in aerospace industry.[3] The alloys were hardened by milling them with low percentage of metal oxide such as Al_2O_3. Another possible strategy consists of milling aluminum or Al-based alloys with the metal oxide (*e.g.*, rutile TiO_2) giving rise to the formation of Al_2O_3 particles in aluminum-based alloys according to the reduction reaction (eq. (3.7)):

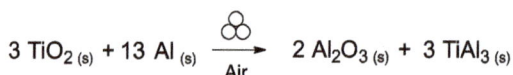

$$3\ TiO_{2\,(s)} + 13\ Al_{(s)} \xrightarrow[\text{Air}]{} 2\ Al_2O_{3\,(s)} + 3\ TiAl_{3\,(s)}$$

Equation 3.7. Al_2O_3 –$TiAl_3$ nanocomposite via mechanochemistry.

The same concept has been profited for the preparation of ceramic nanocomposites by milling a metal and alumina, silica or titania to obtain a fine dispersion of metal NPs in a ceramic matrix.[3] Noteworthy example is the synthesis of Cu-dispersed

Al_2O_3 nanocomposites by BM and further pulsed electric current sintering.[41] Similar studies concern the preparation of Cr (or Nb) alumina composites, obtained by milling equivalent volumes of Al_2O_3 and the respective metal.[42] The preparation of magnetite–SiO_2 nanocomposites according to eq. (3.8) (and Table 3.7) have been also reported:[43]

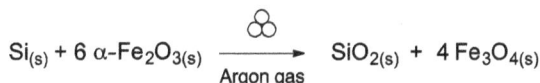

$$Si_{(s)} + 6\ \alpha\text{-}Fe_2O_{3(s)} \xrightarrow[\text{Argon gas}]{} SiO_{2(s)} + 4\ Fe_3O_{4(s)}$$

Equation 3.8. Magnetite–SiO_2 nanocomposite via mechanochemistry. Reaction conditions for method: Si (1 mmol), α-Fe_2O_3 (hematite) (6 mmol).

Table 3.7: Process parameters for mechanosynthesis of Fe_3O_4–SiO_2 nanocomposite.

Parameters of the milling process	Method
Type of mill	SPEX mixer mill
Jar volume (mL)	60
Jar material	SS[a]
Number of balls	67
Balls diameter (mm)	6.35
Weight of each ball (g)	1.05
Milling speed	60 Hz (1060 cpm)
Milling time (min)	264
Milling and resting period (min)	5 and 5
Grinding additive	–
Reaction atmosphere	Ar
Reaction scale (mmol)	6

[a]SS, stainless steel.

The magnetite–SiO_2 nanocomposite is an intermediate of a Fe–SiO_2 nanocomposite obtained in prolonged (43 h) and harsher milling condition (290 rpm) in a planetary ball mill [eq. (3.9) and Table 3.8]:[26]

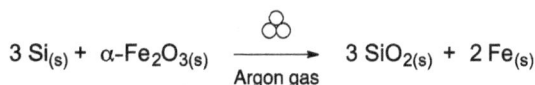

$$3\ Si_{(s)} + \alpha\text{-}Fe_2O_{3(s)} \xrightarrow[\text{Argon gas}]{} 3\ SiO_{2(s)} + 2\ Fe_{(s)}$$

Equation 3.9: Fe–SiO_2 nanocomposite via mechanochemistry. Reaction conditions for method: Si (246 mmol), α-Fe_2O_3 (hematite) (82 mmol).

The average crystallite size of iron in the $Fe-SiO_2$ nanocomposite is about 15 nm with 0.37% strain. About 92% of iron crystallite have dimensions ranging from 5 to 10 nm and about 8% have the size up to 60 nm (Table 3.8).

Table 3.8: Process parameters for the mechanochemical preparation of the $Fe-SiO_2$ nanocomposite.

Parameters of the milling process	Method
Type of mill	Planetary mill
Jar volume (mL)	250
Jar material	SS[a]
Number of balls	50
Balls diameter (mm)	8
Weight of each ball (g)	4.08
Milling speed	290 rpm
Milling time (h)	43
Milling and resting period (min)	5 and 5
Grinding additive	–
Reaction atmosphere	Ar
Reaction scale (mmol)	246

[a]SS, stainless steel.

Recently, a renewing of ball milling has been proposed to obtain novel drug SiO_2 and natural extract SiO_2 nanocomposites for biomedical applications.[44, 45] It represents a new and viable pathway to develop novel pharmaceuticals and cosmetics. It consists in the fine dispersion of drugs or natural extracts in a porous silica matrix, which acts as a controlled drug delivery system. In a typical synthesis, 1 g of fumed silica and ethanolic extract of *Vitis vinifera* (*Vv*) was sealed in an agate vial (Table 3.9). The composition of the starting mixture was selected in order to obtain a dry extract content of *Vv* equal to 1.0, 9.0 and 33.0 wt.%. In the example reported here, it is recommended to use 1.0 wt.% *Vv* content.

After the milling process, samples are dried at room temperature for 48 h in order to eliminate the ethanol of the ethanolic extract.

Finally, it is noteworthy to mention the use of mechanochemistry for noble-metal activation and recycling of elementary palladium and gold. Such process leads to a direct, efficient and one-pot conversion of the metals into water-soluble salts or metal–organic catalysts.[46]

Table 3.9: Process parameters for mechanochemical synthesis of *Vitis vinifera* extract-SiO_2 nanocomposites.

Parameters of the milling process	Method
Type of mill	Planetary mill
Jar volume (mL)	60
Jar material	agate
Number of balls	2
Balls diameter (mm)	20
Weight of each ball (g)	11.3
Milling speed	200 rpm
Milling time (h)	1
Milling and resting period (min)	5 and 5
Grinding additive	–
Reaction atmosphere	Air
Reaction scale (mmol)	16.6

Conclusion

This chapter is an introduction to most of the achievements of mechanochemistry in the field of inorganic chemistry. We can proudly affirm that this is the oldest deepest routed area of mechanochemical synthesis, and its application to the development of inorganic compounds has allowed the preparation of a wide variety of materials by reduced times, increased yields, environmentally friendly and cost-efficient processes.

Therefore, an obvious question comes to mind: "Is the time for inorganic mechanochemistry over?" The answer to this question is: "Absolutely not!" On one side, despite many impressive and valuable materials that have been already obtained, the synthetic mechanisms still need to be explained. On the other side, novel beautiful materials covering a wide range of applications are waiting to be developed.

This is the time to grind and all of you will be very welcome to contribute to a "new Era" of inorganic mechanochemistry.

References

[1] Faraday, M., Lit, Q. J. S. Arts 1820, 8, 374.

[2] James, S. L., Adams, C. J., Bolm, C., Braga, D., Collier, P., Jones, W., Krebs, A., Mack, J., Maini, L., Orpen, A. G., et al. Mechanochemistry: Opportunities for New and Cleaner Synthesis. Chem. Soc. Rev. 2012, 41, 413–447.

[3] Baláž, P., Achimovičová, M., Baláž, M., Billik, M., Criado, M., Delogu, F., Cherkezova-zheleva, Z., Gaffet, E., Jose, F., Dutkova, E., et al. Hallmarks of mechanochemistry: from nanoparticles to technology. Chem. Soc. Rev. 2013, 42, 7571–7637.

[4] Le, M. T., Kim, D. J., Kim, C. G., Sohn, J. S., Lee, J. Nanoparticles of silver powder obtained by mechano-chemical process. J. Exp.Nanosci. 2008, 3, 223–228.

[5] Keskinen, J., Ruuskanen, P., Karttunen, M., Hannula, S. P. Synthesis of silver powder using a mechanochemical process. Appl. Organomet. Chem. 2001, 15, 393–395.

[6] West, A. R. Basic Solid State Chemistry. Wiley, 1996.

[7] Koch, C., Cahn, R. W., Eds. Materials Science and Technology-A Comprehensive Treatment, Processing of Metals and Alloys. VCH Verlagsgesellschaft 1991, 15, 193–245.

[8] Schwarz, R. B., Koch, C. C. Formation of amorphous alloys by the mechanical alloying of crystalline powders of pure metals and powders of intermetallics. Appl. Phys. Lett. 1986, 49, 146–148.

[9] Arceo, L. D. B., Cruz-Rivera, J. J., Cabanas-Moreno, J. G., Tsuchiya, K., Umemoto, M., Calderson, H. Characterization of Cu-Co Alloys Produced by Mechanosynthesis and Spark Plasma Sintering. Mater. Sci. Forum. 2000, 343–346, 641–646.

[10] Karolus, M., Jartych, E., Oleszak, D. Structure and magnetic properties of nanocrystalline Fe-Mo alloys prepared by mechanosynthesis. Acta Phys. Pol. A 2002, 102, 253–258.

[11] Kim, K. J., Sumiyama, K., Suzuki, K. Martensitic transformation of mechanically ground γ'-Fe$_4$N. J. Magn. Magn. Mater. 1995, 140–144, 49–50.

[12] Kim, M. S., Koch, C. C. Structural development during mechanical alloying of crystalline niobium and tin powders. J. Appl. Phys. 1987, 62, 3450–3453.

[13] Tianen, T. J., Schwarz, R. B. Synthesis and characterization of mechanically alloyed Ni-Sn powders. J. Less-Common Met. 1988, 140, 99–112.

[14] Ennas, G., Magini, M., Padella, F., Pompa, F., Vittori, M. On the Formation of Pd$_3$Si by Mechanical Alloying Solid State Reaction. J. Non-Cryst. Solids 1989, 110, 69–73.

[15] Lopez-Baez, I., Martınez-Franco, E., Zoz, H., Trapaga-Martinez, L. G. Structural evolution of Ni-20Cr alloy during ball milling of elemental powders. Rev. Mex. Fis. 2011, 57, 176–183.

[16] Suryanarayana, C. Mechanical Alloying and Milling. Prog. Mater. Sci. 2001, 46, 1–184.

[17] Webber, A. W., Haag, W. J., Wester, A. J. H., Bakker, H. Differences in the amorphization reaction by mechanical alloying of Ni-Zr resulting from different ball-milling techniques. J. Less-Common Met. 1988, 140, 119–127.

[18] Eckert, J., Schultz, L., Hellstern, E., Urban, K. Glass-forming range in mechanically alloyed Ni-Zr and the influence of the milling intensity. J. Appl. Phys. 1988, 64, 3224–3228.

[19] Gaffet, E. Dynamic equilibrium induced by ball milling in the NiZr system. M. Mat. Sci. Eng. A 1989, 119, 185–197.

[20] Ennas, G., Magini, M., Padella, F., Susini, P., Boffitto, G., Licheri, G. Preparation of amorphous Fe-Zr alloys by mechanical alloying and melt spinning method. Part I -. a structural comparison. J. Mater. Sci. 1989, 24, 3053–3058.

[21] Burgio, N., Iasonna, A., Magini, M., Martelli, S., Padella, F. Mechanical alloying of the Fe-Zr system. Correlation between Input Energy and End Products. NUOVO Cim. 1991, 13, 20–23.

[22] Kim, K., Sumiyama, K., Suzuki, K. Ferromagnetic Ar-Mn type Mn-Al alloys produced by mechanochemical methods. J. Magn. Magn. Mater. 1995, 140-144, 49–50.

[23] Enzo, S., Schiffini, L., Battezzati, L., Cocco, G. RDF and EXAFS structural study of mechanically alloyed $Ni_{50}Ti_{50}$. J. Less-Common Met. 1988, 140, 129–137.

[24] Baricco, M., Cowlam, N., Schiffini, L., Macrí, P. P., Copper – Cobalt f . c . c . Metastable Phase Prepared by Mechanical Alloying. 2006, No. June 2015, 37–41.

[25] Corrias, A., Ennas, G., Marongiu, G., Musinu, A., Paschina, G. Influence of boron content of the amorphization rate of Co-B mixtures by mechanical alloying. J. Mater. Res. 1993, 8, 1327–1333.

[26] Corrias, A., Ennas, G., Musinu, A., Paschina, G., Zedda, D. Iron-silica and nickel-silica nanocomposites prepared by high energy ball milling. J. Mater. Res. 1997, 12, 2767–2772.

[27] Magini, M., Basili, N., Burgio, N., Ennas, G., Martelli, S., Padella, F., Paradiso, E., Susini, P. Mechanical alloying of the Pd-Si system. investigation of the early and late milling stages. Mat. Sci. Eng. 1991, A134, 1406–1409.

[28] Kubalova, L. M., Fadeeva, V. I., Sviridov, I. A., Fedotov, S. A. The synthesis of nanocrystalline $Ni_{75}Nb_{12}B_{13}$ alloys by high energy ball milling of elemental components. J. Alloy. Compd. 2009, 483, 86–88.

[29] Oliker, V. E., Sirovatka, V. L., Gridasova, T. Y., Timofeeva, I. I., Bykov, A. I. Mechanochemical synthesis and structure of Ti-Al-B-based alloys. Powder Met. Met. Ceram. 2008, 47, 546–556.

[30] Medina, M. H. Structural and magnetic properties of $Fe_{0.45}Mn_{0.25}Al_{0.30}$ alloys prepared by mechanical alloying. Phys. Stat. Sol. B 2006, 243, 1390–1399.

[31] Mushove, T., Chikwanda, H., Machio, C., Ndlovu, S. Ti-Mg alloy powder synthesis via mechanochemical reduction of TiO_2 by elemental magnesium. Mater. Sci. Forum. 2009, 618–619, 517–520.

[32] Farahbakhsh, S., Tabaian, H., Vahdati, J. Production of nano leaded brass alloy by oxide materials. Adv. Mater. Res. 2010, 83–86, 36–40.

[33] Padella, F., Alvani, C., La Barbera, A., Ennas, G., Liberatore, R., Varsano, F. Mechanosynthesis and Process Characterization of Nanostructured Manganese Ferrite. Mater. Chem. Phys. 2005, 90, 172–177.

[34] Cristobal, A. A., Aglietti, E. F., Conconi, M. S., Porto Lopez, J. M. Structural alterations during mechanochemical activation of a titanium–magnetite mixture. Mater. Chem. Phys. 2008, 111, 341–345.

[35] Sepelak, V. Mechanosynthesis of Nanocrystalline Iron Germanate Fe_2GeO_4 with a Nonequilibrium Cation Distribution. Rev. Adv. Mater. Sci. 2008, 18, 349–352.

[36] Botta, P. M., Aglietti, E. F., Porto-Lopez, J. M. Kinetic Study of $ZnFe_2O_4$ Formation from Mechanochemically Activated Zn – Fe_2O_3 Mixtures. Mater. Res. Bull. 2006, 41, 714–723.

[37] Botta, P. M., Aglietti, E. F., Porto Lopez, J. M. Mechanochemical synthesis of hercynite. Mater. Chem. Phys. 2002, 76, 104–109.

[38] Bellusci, M., Aliotta, C., Fiorani, D., Barbera, A. L., Padella, F., Peddis, D., Pilloni, M., Secci, D. Manganese iron oxide superparamagnetic powder by mechanochemical processing. Nanoparticles functionalization and dispersion in a nanofluid. J. Nanopart. Res. 2012, 14, 904–914.

[39] Rojac, T., Kosec, M., Malic, B., Holc, J. The mechanochemical synthesis of $NaNbO_3$ using different ball-impact energies. J. Am. Ceram. Soc. 2008, 91, 1559–1565.

[40] Briančin, P., Baláž, P., Baláž, E., Boldižárová, J. Preparation, mechanochemical route for sulphide nanoparticles. Mater. Lett. 2003, 7, 1585–1589.

[41] Kim, Y. D., Oh, S.-T., Min, K. H., Jeon, H., Moon, I.-H. Synthesis of Cu dispersed Al_2O_3 nanocomposites by high energy ball milling and pulse electric current sintering. Scr. Mater. 2001, 44, 293–297.

[42] De la Torre, S. D., Garcia, D. E., Claussen, N., Nishikawa, R. Y., Miyamoto, H., Martinez-Sanchez, R., Garcia-Luna, A., Rios-Jara, D. Spark plasma sintering of alumina–Cr and –Nb composites. Mater. Sci. Forum. 2002, 386–388, 299–302.

[43] Scano, A., Cabras, V., Marongiu, F., Peddis, D., Pilloni, M., Ennas, G. New Opportunities in the Preparation of Nanocomposites from Biomedical Applications: Revised Mechanosynthesis of Magnetite-Silica Nanocomposites. Mater. Res. Express. 2017, 4, 025004, Doi: 10.1088/2053-1591/aa5867.

[44] Pilloni, M., Ennas, G., Casu, M., Fadda, A. M., Frongia, F., Marongiu, F., Sanna, R., Scano, A., Valenti, D., Sinico, C. Drug silica nanocomposite: preparation, characterization and skin permeation studies. Pharm. Dev. Technol. 2013, 18, 626–633.

[45] Scano, A., Ebau, F., Manca, M. L., Cabras, V., Marincola, F. C., Manconi, M., Pilloni, M. Novel drug delivery systems for natural extracts : the case study of vitis vinifera extract-SiO_2 nanocomposites. Int. J. Pharm. 2018, 551, 84–96.

[46] Do, J.-L., Tan, D., Friščić, T. Noble metals oxidative mechanochemistry : direct, room-temperature, solvent- free conversion of palladium and gold metals into soluble salts and coordination complexes. Angew. Chem. Int. Ed. 2018, 57, 2667–2671.

Evelina Colacino, Andrea Porcheddu

4 Introducing practical organic mechanochemistry into undergraduate curricula

Making green chemistry an integral part of chemistry education[1] complies with the increasing demand for a sustainable manufacture of chemical products, through the development of cleaner, safer and more efficient chemical transformations *benign by design*.[2] A way to fulfill this objective is to increase the awareness of both students and instructors at each educational level, about the importance of *thinking chemistry differently*[3] when designing a synthesis or a chemical process and to inspire the next generation of scientists and engineers through an engaging learning environment. However, adoption of a sustainable thinking concerns also politics, economists, business leaders and stakeholders.

Introducing students to mechanochemistry and ball milling as a highly successful approach and an alternative to conventional solvent-based procedures contributes to this change in mentality, by questioning about the role and the need for solvents in chemical transformations.

Despite the barriers to its wider adoption that mechanochemistry is surprisingly facing, or exactly because of these barriers, through the selected examples described in this chapter, we aim to (i) encourage the instructors to adopt mechanochemistry and ball-milling in teaching synthetic chemistry, (ii) familiarize undergraduate students to the potential and the practice of mechanochemistry and (iii) contribute to the development of a contemporary approach to preparative organic chemistry, already at the undergraduate level.

The multidisciplinarity of mechanochemistry, touching chemistry, materials science and environmental sciences, as well as the continuous discovery of new areas of application, cannot be covered here exhaustively and we invite interested readers to refer to specialized reviews or books.[4–8]

The examples illustrated in this chapter are selected among the synthetic transformation finding broad application in several areas of chemistry and belong to the general background of an organic chemist. The examples here described are classified in two categories: metal-catalyzed (e.g., coupling reactions, *click* synthesis and olefin metathesis) and metal-free (e.g., condensation, cycloaddition and multicomponent reactions) reactions and include selected examples for the preparation of active pharmaceutical ingredients (API). Mechanochemical ball-milling reactions are here represented using the formalism of three circles in a triangular arrangement (⨂), a symbol recently introduced by Hanusa.[9]

In addition, the following criteria were taken into account when selecting experiments: (i) use of cheap and commercially available reactants, (ii) fit into an average

https://doi.org/10.1515/9783110608335-004

timeframe of an organic chemistry laboratory practices (4–6 h), (iii) leading to full conversion of substrates with high selectivity, (iv) recovery of the product by a straightforward work-up procedure (e.g., the final products being recovered by simply scratching them out of the milling jars, by precipitation/filtration steps in water or in exceptional cases by column chromatography[10]), (v) for the same reaction, use of different grinding materials (agate, stainless steel, zirconium oxide and tungsten carbide), additives[11] and mechanochemical tools (mortar and pestle, mixer- or planetary mill),[12] providing a different type of mechanical stress, (vi) when possible, use of eco-friendly reagents and grinding additives and (vii) use of green chemistry metrics to assess the sustainability of the process.

Implementation of mechanochemistry to conduct well-known transformations fulfills several other objectives. The students (i) are encouraged to review their basic knowledge while learning a new approach to synthetic chemistry, based on solvent-free procedures involving emerging and enabling technologies, (ii) increase their sustainable thinking by increasing their awareness about the environmental impact of a synthetic process and laboratory operations, (iii) think outside the box through the comparison of the green chemistry metrics for both mechanochemical and solvent-based procedures and (iv) learn how to optimize a mechanochemical reaction and how the chosen process parameters influence mechanochemical reactivity and the outcome of a milling experiment.

We sincerely hope that the described examples will contribute to drive the right behavior of both students and instructors in chemistry laboratories.

4.1 Metal-catalyzed organic reactions

4.1.1 Selected experiments

4.1.1.1 Suzuki–Miyaura cross-coupling

The formation of C–C bonds via Suzuki–Miyaura cross-coupling between aryl halides with organoboron reagents has found wide applications in both academic and industrial settings. Since the first report[13] of mechanochemically activated solvent-free reaction using a planetary ball mill, several improvements were proposed. Despite the fact that various possibilities have been reported with excellent degrees of performances, selected examples suitable for undergraduate level are herein presented (Scheme 4.1).

Methods A and B do not require any ligand, however, their application is limited compared to method C, broadly applicable for the palladium-catalyzed cross-coupling reaction of a wide range of aryl halide/arylboronic acid combinations, including aryl chlorides (which are usually poorly reactive in solution).

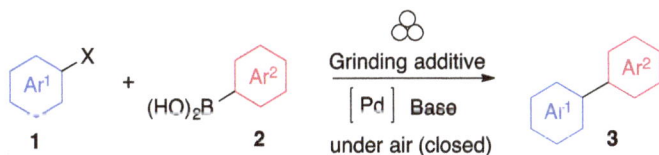

Scheme 4.1: Suzuki–Miyaura reaction by mechanochemistry. Reaction conditions for method **A**[14]: X = Br (mainly), **1** (1 mmol), **2** (1 mmol), Pd(OAc)$_2$ (4.5 mol%), Et$_3$N (3.0 mmol), NaCl (2.0 g) (**3**, 10 examples, 33–95% yield); method **B**[15–16]: X = I, Br (mainly), **1** (5 mmol), **2** (6.19 mmol), Pd(OAc)$_2$ (3.56 mol%), KF-Al$_2$O$_3$ (5 g, 32 wt.% of KF) (**3**, 15 examples, 18–98% yield); method **C**[17]: X = I, Br, Cl, **1** (0.3 mmol), **2** (0.36 mmol), Pd(OAc)$_2$ (3.0 mol%), DavePhos (4.5 mol%), CsF (0.9 mmol), H$_2$O (1.1 mmol), 1,5-cod (0.12 µL/mg) (**3**, 33 examples, 43–99% yield).

The use of the grinding agent sodium chloride (method A) [14] is necessary to prevent the formation of sticky mixtures provoking unproductive milling. Chemically inert, cheap and with low toxicity, sodium chloride keeps the mixture in the form of a free-flowing powder avoiding that very soft or waxy substances stick to the milling jar walls.

For method B,[15–16] the addition of a sacrificial base, necessary to the catalytic cycle, is not needed, being the base (KOH) generated in situ[15] by decomposition of the inorganic support KF-Al$_2$O$_3$ with water.[15]

More recently, a general and scalable protocol for solid-state Suzuki–Miyaura was reported under liquid-assisted grinding conditions (LAG, method C).[17] Enhanced conversion rates are achieved also for challenging substrate combinations by addition of a small amount of olefins (typically 1,5-ciclooctadiene, 1,5-cod), acting as both dispersants of the palladium-based catalyst and stabilizer for the active catalytic Pd(0) species.

The technical variables of the used apparatus influence the reaction outcome (and not only the reaction conditions such as the amount of substrate and catalyst) (Table 4.1).

However, whatever is the mechanochemical method selected for the palladium-catalyzed solid-state Suzuki–Miyaura cross-coupling, compared to procedures in solution, mechanochemical activation presents several advantages. It is (i) technically and operationally simple, (ii) reaction conditions are simplified (not requiring oxygen-free atmosphere), (iii) safety is improved (no need of solvent or heating), (iv) high yielding, and it provides (v) better energy balance (especially compared to microwave-assisted reactions) and (vi) faster kinetics.

Table 4.1: Process parameters for mechanochemical Suzuki–Miyaura reaction.

Key parameters of the milling process	Method		
	A[14]	B[15, 16]	C[a,17]
Type of mill	Mixer mill	Planetary mill	Mixer mill
Jar volume (mL)	10	45	1.5 (25)
Jar material	SS[b]	SS[b]	SS[b]
Number of balls	1	6	1 (6)
Balls diameter (mm)	n.r.[c]	15	5 (10)
Weight of each ball	n.r.[c]	n.r.[c]	n.r.[c]
Milling speed	30 Hz	800 rpm	25 Hz
Milling time (min)	10	15	99
Grinding additive	NaCl	–	H_2O and 1,5-cod
Reaction scale (mmol)	1	5	0.3 (8)

[a]The process parameters for the gram-scale preparation (*ca.* 2.0 g) are given in parenthesis;
[b]SS, stainless steel;
[c]n.r., not reported.

4.1.1.2 Mizoroki–Heck cross-coupling

After the first report in 2004,[18] the mechanochemical Heck cross-coupling was further investigated,[19–21] and its oxidative version was also reported.[22–23] Silica gel (method A)[21] or solid poly(ethylene)glycol PEG-2000-OH (method B)[20] were used as grinding auxiliaries (Scheme 4.2 and Table 4.2). In the case of (*E*)-stilbenes **4**, silica gel was also an adsorbent in the reaction. Both methods present the advantage of avoiding the use of phosphine ligands, usually necessary to stabilize palladium active species. Indeed, the metal colloids generated in situ are stabilized by the presence of tetrabutyl ammonium bromide (TBAB) and PEG-2000-OH, respectively.[24]

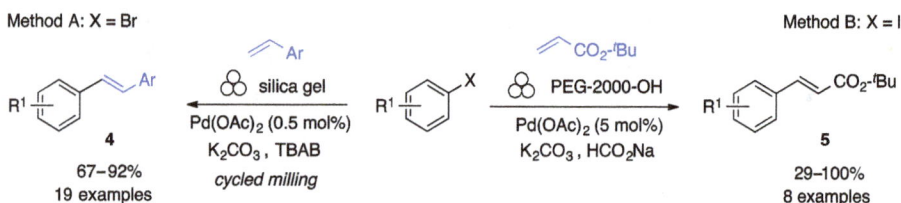

Scheme 4.2: Mizoroki–Heck reaction by mechanochemistry.

Table 4.2: Process parameters for mechanochemical Mizoroki–Heck reaction.

Parameters of the milling process	Method	
	A[21]	B[20]
Type of mill	Planetary mill	Mixer mill
Jar volume (mL)	105	10
Jar material	SS[a]	SS[a]
Number of balls	n.r.[b]	2
Balls diameter (mm)	8	7
Weight of each ball (g)	234 [c]	n.r.[b]
Milling speed	1290 rpm	30 Hz
Milling time (min)	45–60	60
Milling and resting period (min)	15 and 5	/
Grinding additive	Silica gel	PEG-2000-OH
Reaction scale (mmol)	5	0.1

[a]SS, stainless steel;
[b]n.r., not reported;
[c]the total weight of the balls is given.

However, when using a PEG matrix, the activation of the precatalysts $Pd(OAc)_2$ required the use of an external reducing agent such as sodium formate. A notable advantage of PEG-based procedure relies on the straightforward separation of the cinnamates **5** from the metallic catalytic system via a precipitation/filtration step,[24] even if the scope of the reaction is limited (reactions with aryl chloride and bromide were unreactive).[20]

Methods A and B share common benefits, which include (i) the relatively low catalysts loading without using expensive ligands (and avoiding toxic phosphines); (ii) high selectivities and yields in short reaction times; (iii) simple experimental set-up, simplified and safer reaction conditions (the use of toxic solvents is avoided and an inert atmosphere is not required).

Worth of notice is the recent preparation of E-stilbenes also by Ru-catalyzed mechanochemical cross-metathesis reaction[25] or by Witting's reaction.[26–29]

4.1.1.3 Sonogashira cross-coupling

Pioneer investigations by Mack's group,[30] and Stolle's[31, 32] demonstrated that ball milling was effective for the Sonogashira cross-coupling of aryl halides and terminal alkynes, also in the absence of copper (Scheme 4.3).

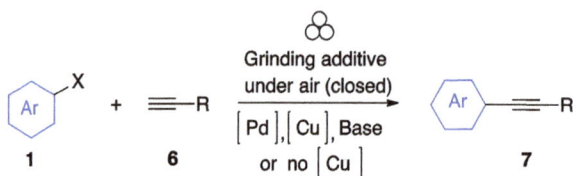

Scheme 4.3: Sonogashira reaction by mechanochemistry (with or without copper sources). Reaction conditions for method A:[30] X = I, Br; R = Ph, TMS; 1 (0.98 mmol), 6 (1.05 mmol), Pd(PPh$_3$)$_2$ (2.5 mol%), CuI (1 mol%), K$_2$CO$_3$ (0.99 mmol) (7, 16 examples, 3–93% yield); methods B1 and B2:[30] X = I, Br; R = TMS; 1 (0.98 mmol), 6 (1.05 mmol), Pd(PPh$_3$)$_2$ (2.5 mol%), "copper source" (jar and ball, Table 4.3), K$_2$CO$_3$ (0.99 mmol) (7, 7 examples, 42–90% yield); method C (copper free):[31, 32] X = I, Br; R = Ph; 1 (2.0 mmol), 6 (2.5 mmol), Pd(OAc)$_2$ (5 mol%), DABCO (2.5 mmol), grinding auxiliary (5 g, Table 4.3) (7, 18 examples, up to 99% conversion of aryl halide 1).

Table 4.3: Process parameters for mechanochemical Sonogashira reaction.

Parameters of the milling process	Method		
	A[30]	B1 / B2[30]	C[31, 32]
Type of mill	SPEX mixer mill	SPEX mixer mill	Planetary mill
Jar volume (mL or in)	2.0 x 0.5[a]	2.0 x 0.5[a]	45
Jar material	SS[b]	SS[a]/ Cu	Agate or ZrO$_2$
Ball material	W/C[b]	Cu / Cu	Agate or ZrO$_2$
Number of balls	1	1	6
Balls diameter (mm or in)	0.250 in[a]	3/32 in	15
Weight of each ball	n.r.[c]	n.r.[c]	n.r.[c]
Milling speed	18 Hz	18 Hz	800 rpm
Milling time (min or h)	17 h	17 h	15 min
Grinding additive	/	/	SiO$_2$/α-Al$_2$O$_3$
Reaction scale (mmol)	1	1	2.5

[a]Custom-made 2.0 × 0.5 in screw-capped stainless steel vial filled with a 0.250 in ball;
[b]SS, stainless steel; W/C, tungsten carbide;
[c]n.r., not reported.

To make this reaction more environmentally benign is not only the elimination of the reaction solvent and the need of inert conditions, but also the use of the jar material (and balls) as "catalyst," an alternative *copper source* to CuI, leading to similar results (method A *vs.* methods B1 and B2 and Table 4.3).[30] Remarkably, *copper-free* Sonogashira reaction was also reported in stainless steel jars,[30] agate or zirconium oxide[31, 32] (Table 4.3). The better results were obtained with 1,4-diazabicyclo[2.2.2]-octane (DABCO) as base in the presence of grinding additives as SiO_2 (fused quartz sand) or basic α-Al_2O_3, affording full conversion of substrates and 99% selectivity in only 15 min (Table 4.3).

Side reactions as dehalogenation of aryl halides or homocoupling of alkynes were not observed in this case.

4.1.1.4 Copper-catalyzed 1,3-dipolar cycloaddition reaction (CuAAC "click" reaction)

The copper-catalyzed 1,3-dipolar cycloaddition between alkynes and azides, also known as *click reaction*, is a straightforward method for the stereospecific preparation of 1,4-disubstituted-1,2,3-triazoles **9** in high yields and 100% *atom economy*.[33] Sodium ascorbate is usually required as reducing agents and to prevent the formation of oxidative homocoupling products.

Triazoles are important scaffolds for medicinal chemistry, for the derivatization of biomolecules and in polymer synthesis. Several mechanochemical approaches were disclosed not only to broaden the scope of the reaction, but also to circumvent the risks associated when handling azides. Organic azides are toxic and hazardous in general, in addition, low-molecular-weight azides are also sensitive to shocks (Scheme 4.4).[34–37]

In this regard, in situ preparation of azides from suitable precursors (methods C–E, Scheme 4.4 and Table 4.4) is certainly a safer approach, especially when mechanochemical processes are used as alternative to other solution-based energy inputs. Compared to solution-based procedures, the *click* reaction performed by mechanochemistry performs without any reducing agent (only used with particularly complex substrates)[38] and with no competing homocoupling reaction involving alkynes (Glaser reaction). In this regard, the mechanochemical *click* reaction is more ecofriendly than its solution-based counterpart, providing an improved *reagent* and *waste economy*.

The *click* reaction performed well also with copper powder (in stoichiometric amount, method B).[36] The method is fast and scalable (up to 10 g) and the products are separated from the metal powder by filtration. Similarly, high yields in short reaction times (15 min) were reported with alkyl azides, replacing the copper salts by the copper peeled off the jars (and balls), used to shake the mixture.[34] The *copper catalyst* (the jar) was "recycled" (used) over 100 reactions without decrease in yield and reaction rate (method C, Table 4.4).[34] However, the same approach was unsatisfactory in term of kinetics (16 h) for the three-component reaction described in method C[34] (Scheme 4.4 and Table 4.4), probably due to the slow nature of alkyl azide formation in situ.

Method A,B   Methods C,D

(Cycled milling)

Method E:

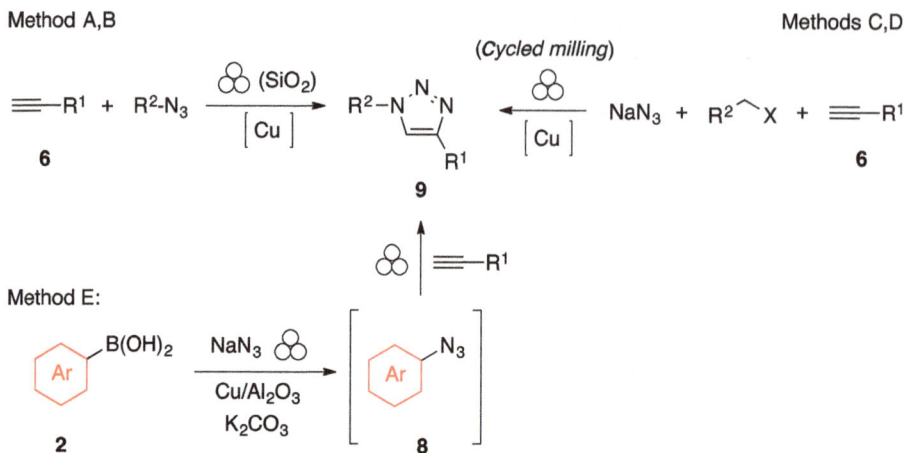

Scheme 4.4: Mechanochemical *click* reaction to 1,4-disubstituted triazoles. Reaction conditions for method **A**:[37] R^1, R^2 = Ph, aryl, alkyl, **6** (1.0 mmol), R_2-N_3 (1.1 mmol), Cu(OAc)$_2$ (5 mol%), grinding auxiliary SiO$_2$ (5 g, Table 4.4), (**9**, 17 examples, up to 99% GC yield); method **B**:[36] R^1 = Ph, alkyl; R^2 = benzyl, alkyl; **6** (1.0 mmol), R_2-N_3 (1.0 mmol), Cu powder (1 mmol), (**9**, 11 examples, 81–99%); method **C**:[34] X = Br; R^1 = Ph, TMS, alkyl; **6** (1.6 mmol), R_2CH_2X (1.5 mmol), NaN$_3$ (3.0 mmol), "copper source" (jar and ball, Table 4.4), (**9**, 12 examples, 33–95%); method **D**:[35] X = Br, Cl; R^2 = aryl, alkyl; **6** (1.0 mmol), R_2CH_2X (1.0 mmol), NaN$_3$ (1.0 mmol), Cu(II)/Al$_2$O$_3$ (10 mol%), (**9**, 16 examples, 70–96% yield); method **E**:[35] *step 1* **2** (1.0 mmol), NaN$_3$ (3.0 mmol), *step 2* **6** (1.0 mmol), Cu(II)/Al$_2$O$_3$ (10 mol%), (**9**, 7 examples, 83–90%).

Copper(II) sulfate supported on alumina (Cu/Al$_2$O$_3$) allowed the high-yielding and fast mechanochemical synthesis of triazole derivatives **9** from terminal alkynes and alkyl azides prepared in situ (method D, Scheme 4.4 and Table 4.4).[35] However, aryl azides **8** could not be obtained *via* this method, but better through a *one-pot*/two step synthesis from boronic acids **2** (method E).[35] Catalyst recycling was possible up to eight runs without loss of activity.

For all methods A–E, the final products are easily obtained after evaporation of a small amount of an organic solvent (ethyl acetate) used for recovering the product from the jars [34] or by filtrating the support and washing it [with methyl *tert*-butyl ether (MTBE), or ethanol].[35–37]

4.1.1.5 Negishi cross-coupling

The Negishi cross-coupling reaction[39] finds extensive application in the field of total synthesis[40] for the very selective formation of C–C bonds between complex synthetic intermediates. The palladium-catalyzed coupling of organozinc intermediates with organohalides is usually performed at low or room temperature, in an oxygen- and water-free environment, to prevent decomposition of the organozinc compound.

Table 4.4: Process parameters for mechanochemical CuAAC *click* reaction.

Parameters of the milling process	Method			
	A[37]	B[36]	C[34]	D/E[35]
Type of mill	Planetary mill	Planetary mill	SPEX mixer mill	Planetary mill
Jar volume	45 mL	50 mL	2.0 x 0.5[a]	25 mL
Jar material	ZrO_2	SS[b]	Cu	SS[b]
Ball material	ZrO_2	SS[b]	Cu	SS[b]
Number of balls	6	1500/48[b]	1	6
Balls diameter (mm)	15	2/5 [c]	0.250 in[a]	10
Weight of each ball	n.r.[d]	n.r.[d]	n.r.[d]	n.r.[d]
Milling speed	800 rpm	650 rpm	18 Hz	600 rpm
Milling time	10 min	5–30 min	15 min or 16 h	1 h
Milling and resting period	–	–	–	10 min / 5 s
Grinding additive	SiO_2[e]	–	–	Al_2O_3
Reaction scale (mmol)	1	1–10	1.6	1

[a]Custom-made 2.0 × 0.5 in screw-capped stainless steel vial filled with a 0.250 in ball;
[b]SS, stainless steel;
[c]balls of different sizes were used simultaneously;
[d]n.r., not reported;
[e]fused quartz sand was used.

Indeed, organozinc reagents are not commercially available but are prepared immediately prior to their use *via* tedious procedures, success of which is also dependent on the physical form of used zinc (particle sizes, surface area, etc.). These facts have delayed the full development of Negishi cross-coupling compared to other metal mediated cross-coupling reactions, handled in less stringent conditions.

Very recently, a general *one-pot/two-step* mechanochemical method (also compatible with many functional groups) proved to be extremely efficient to activate zinc with the concomitant *in situ* formation of organozinc species, then engaged *one-pot* in a Negishi cross-coupling reaction (Scheme 4.5 and Table 4.5).[41]

It displays a broad substrate scope for both $C(sp^3)$–$C(sp^2)$ and $C(sp^2)$–$C(sp^2)$ couplings and it is operationally simple, being the reaction conducted in air and without the need of complicated experimental setup for dry conditions. Moreover, the physical form of commercially available zinc metal is irrelevant for the preparation of organozinc compounds. The reaction is catalyzed by the commercial Pd-PEPPSI-*i*Pent (R = iso-pentyl, Scheme 4.5) catalyst in the presence of dimethylacetamide

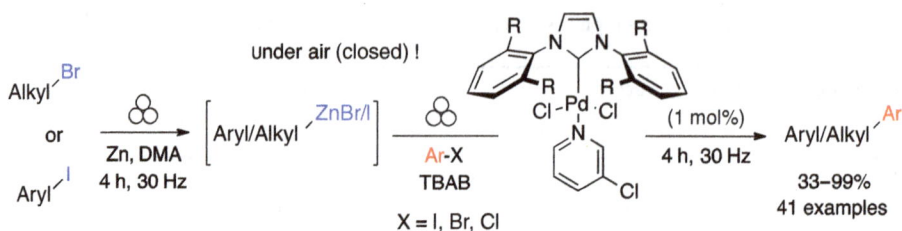

Scheme 4.5: Mechanochemical *one-pot/two-step* reaction for organozinc preparation and Negishi cross-coupling.

Table 4.5: Process parameters for mechanochemical Negishi reaction.

Parameters of the milling process	Method
Type of mill	Mixer mill
Jar volume (mL)	10
Jar material	SS[a]
Number of balls	1
Balls diameter (mm)	n.r.[b]
Weight of each ball (g)	4
Milling speed	30 Hz
Milling time (h)	8
Milling and resting period (min)	One-pot/two-step reaction
Grinding additive	DMA and TBAB
Reaction scale (mmol)	1

[a]SS, stainless steel;
[b]n.r., not reported.

(DMA) and TBAB as grinding auxiliaries. The *one-pot/one-step* protocol Negishi coupling was also effective in the same conditions, mixing all together the reagents, catalysts and grinding auxiliaries and milling for 8 h at 30 Hz.

4.1.1.6 Olefin metathesis

Olefin metathesis[42] is one of the most powerful reaction for the formation of carbon–carbon bonds finding application in several industrial processes. Compared to most of the organic reactions, olefin metathesis creates fewer undesirable by-products and hazardous waste, that are also easily removable (Table 4.6). In this regard, performing cross- and ring-closing metathesis (CM and RCM respectively) by mechanochemistry represents an additional advantage and value-added to the intrinsic simplicity of the reaction (Scheme 4.6).[25]

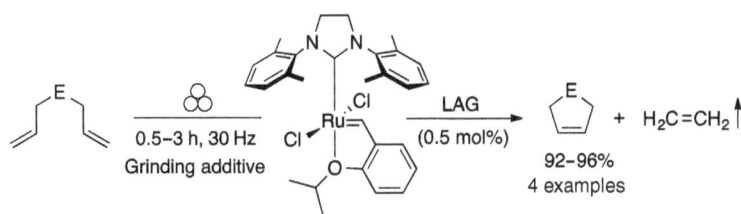

Scheme 4.6: Mechanochemical ring-closing metathesis (RCM).

Table 4.6: Process parameters for mechanochemical olefin metathesis.

Parameters of the milling process	Method
Type of mill	Mixer mill
Jar volume (mL)	10
Jar material	Teflon
Ball material	SS[a] or aluminum[b]
Number of balls	1[b] or 2
Balls diameter (mm)	10 or 20[b]
Weight of each ball (g)	4 or 8[b]
Milling speed	30 Hz
Milling time (h)	0.5–3
Grinding additive	NaCl with EtOAc or PC[c]
Reaction scale (mmol)	1–10

[a]SS, stainless steel;
[b]alternatively, one ball in aluminum was used (20 mm diameter, 8 g);
[c]PC, propylene carbonate.

The reaction outperforms the commercially available second generation Hoveyda–Grubbs catalyst (Hov-II) added portion-wise for which the preparation by mechanochemistry was also recently reported.[43] Combination of a Teflon jar and stainless steel balls was used to perform the reaction, keeping the jar slightly open to allow ethylene formed during the synthesis to escape. Neat grinding was unsuitable, but excellent conversions and yields were obtained in the presence of NaCl as grinding auxiliary alone or in combination with AcOEt or highly polar propylene carbonate (PC), with faster reaction kinetics for liquid substrates. The reaction is 10-fold scalable (up to 3 g) with only 50% increase of the initial amount of the catalyst. The Ru-alkylidene catalyst was quenched by milling for an additional 15–60 min (at 30 Hz) in the presence of a solution of diethylene glycol vinyl ether in dichloromethane. Residual Ru-metal was scavenged with an aqueous solution of L-cysteine or ethylenediaminetetraacetic acid disodium salt (EDTA-Na$_2$) solution and the final products recovered as precipitates.

4.2 Metal-free organic transformations

4.2.1 Selected experiments

4.2.1.1 Wittig reaction

The Wittig olefination involves the reaction of organophosphorus ylides with aldehydes or ketones to give substituted alkenes and triphenylphosphine oxide. The E:Z selectivity depends on the nature (and stability) of the ylide (Scheme 4.7). In solution, moderate to high selectivity for the Z-alkene is observed for unstabilized ylides (R^1 = alkyl), while the E-alkene is the favorite for stabilized ylides (R^1 = ketone, ester). With semistabilized ylides (R^1 = aryl), the E:Z selectivity is often poor.

Mechanochemical activation proved to be very effective not only for the preparation of different type of ylides[27, 29] but also for the one-pot Wittig reaction among a carbonyl compound, triphenyl phosphine (also supported on a polystyrene resin)[26, 28] and an organic halide in the presence of potassium carbonate (Scheme 4.7 and Table 4.7).

Stabilized ylides were generated mechanochemically and isolated in 99% yield, while semistabilized and nonstabilized phosphoranes were generated in situ and trapped one-pot by a solid carbonyl, generating mainly E-alkenes with high selectivity. The Wittig reaction by mechanochemistry presents two major differences compared to the traditional solution-based procedures: (a) straightforward access to nonstabilized phosphoranes, avoiding multi-step synthesis for their preparation and the use of strong bases; (b) a reversed double bond selectivity, leading to more thermodynamically stable E-stilbenes.[27, 29]

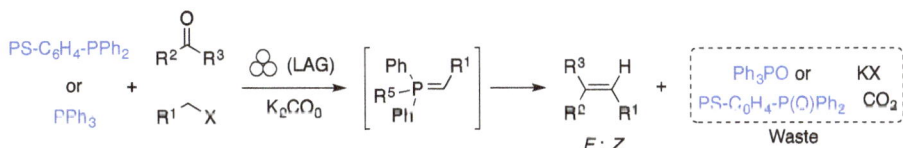

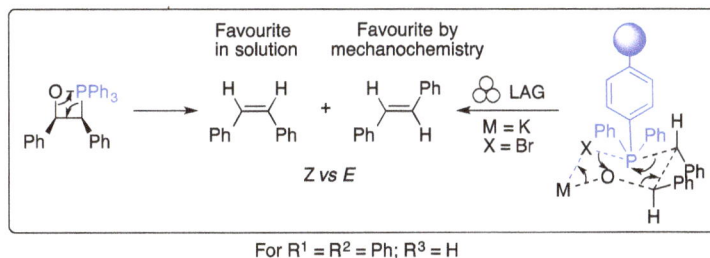

Scheme 4.7: Wittig reaction in a ball mill and investigations on the selectivity.

Table 4.7: Process parameters for mechanochemical Wittig reaction.

Parameters of the milling process	Method	
	A[a, 27, 29]	B[b, 26, 28]
Type of mill	SPEX mixer mill	
Jar volume (mL)	n.r.[c]	3
Jar material	SS or W/C[d]	SS[d]
Number of balls	n.r.[c]	1
Balls diameter (mm)	n.r.[c]	5
Weight of each ball (g)	21 or 70[d]	n.r.[c]
Milling speed	18 Hz	18 Hz
Milling time (h)	3–20	16
Milling and resting period (min)	–	–
Grinding additive	–	LAG
Reaction scale (mmol)	2–7	1

[a]PPh_3 was used;
[b]$PS-C_6H_4-PPh_2$ was used;
[c]n.r., not reported;
[d]SS, stainless steel (21 g), W/C, tungsten carbide (70 g). The total weight of balls is indicated.

The reaction was also investigated by replacing triphenylphosphine by triphenyl-phosphine-functionalized polystyrene (PS-C_6H_4-PPh$_2$).[26, 28] In this case, the elimination of the phosphine oxide derivative is achieved in a chromatography-free manner, by simple filtration. Additionally, it was demonstrated that diasteroselectivity of the reaction is not influenced by electronic effects but it is strongly dependent on the polarity of the solvent used in LAG and the interactions between the reagents participating in the reaction, the proper ion pairing[26] (M and X, Scheme 4.7) and the polystyrene backbone of the resin.

An alternative reaction pathway to the four-membered oxaphosphetane is proposed to occur by milling in neat conditions, generating six-membered rings having the larger groups preferentially in equatorial positions and accounting for the dominant E-selectivity (Scheme 4.7).[26, 28] Moreover, the reactivity and the reaction rate under LAG conditions were increased compared to neat milling. Moreover, the E-selectivity is favored in neat conditions or with LAG experiments in apolar milling environment (e.g., ethyl acetate, n-hexane and toluene), while Z-selectivity is observed in more polar LAG conditions (e.g., ethanol, isopropanol and dichloromethane).

The neat mechanochemical Horner–Wadsworth–Emmons (HWE) version of the Wittig reaction involving the use of more stabilized phosphonate was also reported, leading to high diasteroselectivity in favor of the Z-alkene.[44]

4.2.1.2 (Hetero)-Diels–Alder reaction

The Diels–Alder reaction involving the pericyclic reaction between a conjugate diene and a substituted alkene (dienophile) allows the introduction of chemical complexity in the total synthesis of natural product and new materials. The simultaneous construction of two new carbon–carbon bonds occurs with a good control of both regio- and stereoselectivity. When the dienophile is a π–system involving heteroatoms (e.g., imine), the reaction is named (hetero)-Diels–Alder, generating the corresponding heterocycles (Scheme 4.8).

The investigation of (hetero)-Diels–Alder reaction by mechanochemistry allowed (Scheme 4.8 and Table 4.8) to disclose some general trends and advantages compared to the solution-based procedures. Generally speaking, the process conditions (milling speed, milling media and number of balls) greatly influenced the conversion (and yields) of the reaction. The optimum results were obtained at higher milling speed,[45] using harder (and denser) media such as stainless steel[49] or adding grinding agents such as thymol.[47, 48] In this last case, the solid state reactivity is fostered by the formation of an eutectic mixture with benzoquinone (method C).[48] Moreover, differently than solution-based procedures, by mechanochemistry: (a) the reactions were always highly chemo- and diastereoselective (the endo- or cis-products were formed exclusively), (b) the reagent ratio was improved, avoiding the use of a large excess of dienophile (reagent economy) and post-reaction work-up and purification; (c) the operational setup is simplified (no need to exclude moisture); (d) faster kinetics at room temperature (no thermal activation was needed).

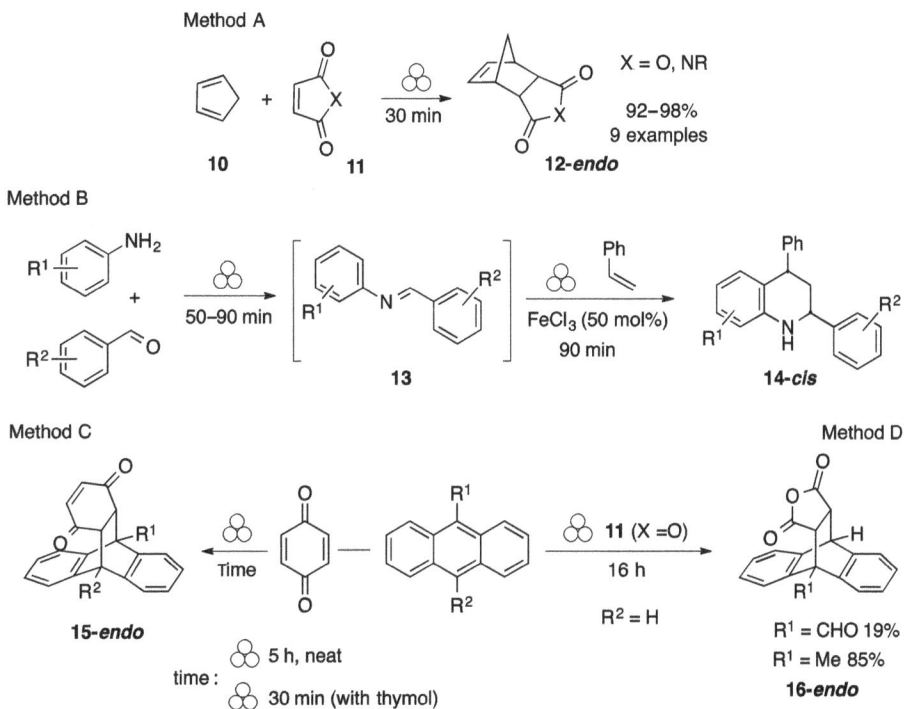

Scheme 4.8: Mechanochemical (hetero)-Diels–Alder reaction. Method A:[45] **10** (5.2 mmol), **11** (5.0 mmol), (**12-endo**, 9 examples, 92–98% yield); method **B**:[46] R^1 = H, Me, OMe, Cl; R^2 = Me, NO_2, Cl; step 1: Ar-CHO (2.0 mmol), Ar-NH_2 (2.0 mmol), step 2: styrene (2.5 mmol), $FeCl_3$ (50 mol%), (**14-cis**, 19 examples, 71–91%); method **C**:[47, 48] R^2 = R^3 = H, Br, alkyl; antracene derivative (1.0 mmol), benzoquinone (1.0 mmol), with or without thymol as the grinding additive (10 mmol%), (**15-endo**, 3 examples); method **D**:[49] antracene derivative (0.5 mmol), **11** (X = O, 0.5 mmol) (**16-endo**, 2 examples, 19 and 85% yield).

Table 4.8: Process parameters for mechanochemical (hetero)-Diels–Alder reaction.

Parameters of the milling process	Method			
	A[45]	B[46]	C[47, 48]	D[49]
Type of mill	Mixer mill	Mixer mill	Automatic mortar	SPEX mixer mill
Jar volume	25 mL	25 mL	n.r.[a]	2.0 x 0.5 in[b]
Jar material	SS[c]	SS[c]	Agate	SS[c]
Ball material	SS[c]	SS[c]	Agate	SS[c]
Number of balls	1	1	1	4
Balls diameter	7 mm	7 mm	n.r.[a]	0.125 in[b]
Weight of each ball	n.r.[a]	n.r.[a]	n.r.[a]	n.r.[a]
Milling speed	30 Hz	30 Hz	50 Hz[d]	18 Hz
Milling time	30 min	Stepwise up to 3h[e]	30 min	16 h
Milling/resting period	–	–	–	–
Grinding additive	–	–	(Thymol)[f]	–
Reaction scale (mmol)	5	2	1	0.5

[a]n.r., not reported;
[b]custom-made 2.0 × 0.5 in screw-capped stainless steel vial filled with a 0.125 in ball;
[c]SS, stainless steel;
[d]the amplitude was 2 mm;
[e]*step 1*: 50–90 min, *step 2*: 90 min;
[f]the experiments were performed with or without thymol.

4.2.1.3 Knoevenagel condensation reaction

Active methylene moiety adds to a carbonyl compound (aldehydes or ketones) to afford conjugated enones and arylidene derivatives, by elimination of one molecule of water. In solution, the Knoevenagel condensation is a base-catalyzed reaction strongly dependent on the solvent used (Table 4.9). By mechanochemistry, stoichiometric amounts of the reactants (malononitrile[50, 51] or barbituric acid derivatives[52]) were milled with no base, leading to 100% yields, and no need of work up (*waste-free reaction*) (Scheme 4.9). The only side product was a molecule of water, easily removed upon heating at 80 °C.

With barbituric acid derivatives, the conversion was always full; however, this can also be due to a concomitant thermal activation, or due to the heat produced by friction during milling. In the case of malononitrile, full conversion was aldehyde-

Method A   Method B

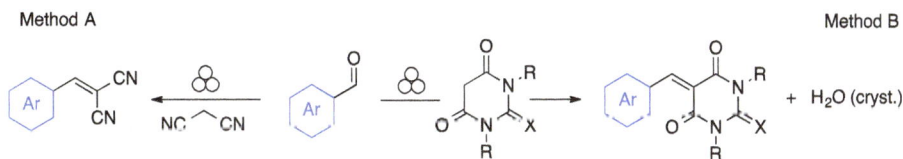

Scheme 4.9: Mechanochemical Knoevenagel condensation reaction (X = O, S).

Table 4.9: Process parameters for mechanochemical Knoevenagel reaction.

Parameters of the milling process	Method		
	A[50]	B[52]	B (large scale) [52]
Type of mill	Planetary mill	Mixer mill	Simoloyer[a]
Jar volume	45 mL	10 mL	2 L
Jar material	SS[b]	SS[b]	SS[b]
Number of balls	5	2	n.r.[c]
Balls diameter (mm)	15	12	5
Weight of balls	n.r.[c]	n.r.[c]	2 kg[a]
Milling speed	800 rpm	25 Hz	1200 rpm
Milling time (h)	1	1	1
Milling/resting period	–	–	–
Grinding additive	–	–	–
Reaction scale	20 mmol	2 mmol	200 g

[a]A Simoloyer CM01-2RM-S1 horizontal high-grade steel ball mill with stellite rotor was used, filled
with 2 Kg of CR6 balls;
[b]SS, stainless steel;
[c]n.r., not reported.

dependent and partially converted pre-milled reaction mixtures reached full conversion to the final product upon aging at room temperature.[50, 51] Additionally, malononitrile and methyl cyanoacetate were milled in the presence of several different
aldehydes, in basic conditions (calcite or fluorite were used), leading exclusively to
the E-alkene.[53] In comparison with solution-based procedures, side reactions as te-
lomerization of substrate RCH_2CN or polymerization of the product E-ene-
nitrile were not observed.

The mechanochemical Knoevenagel reaction was studied by *in situ* monitoring
techniques,[54, 55] its kinetic behavior disclosed[56] and its feasibility at gram-scale

demonstrated (*e.g.*, batches of 200 g).[57–59] Noteworthy, vanillin and barbituric acid demonstrated unusual formation of a cocrystal intermediate between barbituric acid and vanillin before the Knoevenagel condensation reaction.[55] Formation of the cocrystal could be controlled by using different liquid additives. It could also be avoided if a strong base, such as diisopropyethyllamine (DIPEA), was used affording the target condensation product directly.

4.2.1.4 Multicomponent reactions

The formation of several bonds in a single operation by combining together (simultaneously of by sequential addition) three or more starting materials is referred to as multicomponent reactions (MCR). Their high synthetic efficiency allows to reduce the number of synthetic and purification steps, lowering the environmental impact of the synthesis (*waste reduction*), in comparison with more conventional linear, multistep synthetic strategies. In combination with mechanochemical activation, MCR has additional advantages such as improved yields in shorter reaction times, with a diminished generation of solvents, especially when purification by column chromatography can be avoided. Several examples of mechanochemical MCRs were already described.[60]

The Biginelli three-component reaction (3-CR, Scheme 4.10) was studied and excellent results were obtained by grinding stoichiometric amounts of the reactants manually in a mortar[61, 62] or by using a food mixer.[63] However, to achieve reproducible results, dedicated milling equipment needs to be employed.[64–66] The mechanochemical preparation of several 3,4-dihydropyrimidine derivatives occurred in most of the cases in the presence of a Lewis or Brønsted acid catalyst (such as *p*-toluenesulfonic acid, ammonium acetate, ZnO nanoparticles or $SnCl_4$). A catalyst-free procedure was reported,[63] the mechanical shocks providing the direct activation of the substrates with full conversion, in a *work-up* and *waste-free* process affording quantitative yields in 30 min (Scheme 4.10 and Table 4.10). However, effects due to the exothermic nature of the Biginelli reaction cannot be excluded.

Isocyanide-based mechanochemical Passerini 3-CR and Ugi 4-CR were also reported using commercially available reactants (Scheme 4.10 and Table 4.10). Considering the importance of isocyanides in MCR and as reagents in drug design, their mechanochemical preparation by Hofmann carbylamine reaction was also proposed.[70] The mechanochemical preparation of valuable Passerini (in neat conditions) and Ugi (by LAG with MeOH) adducts[67] was faster than in solution-based procedures, but unfortunately, purification by column chromatography was necessary. Similar improved kinetics were observed in the mechanochemical preparation of α-amino nitriles and α-amino amides by Strecker synthesis involving HCN surrogates such as KCN[69] or its (preferable) non-toxic alternative $K_3Fe(CN)_6$[68] in the presence of silica as grinding auxiliary.

Biginelli 3-CR

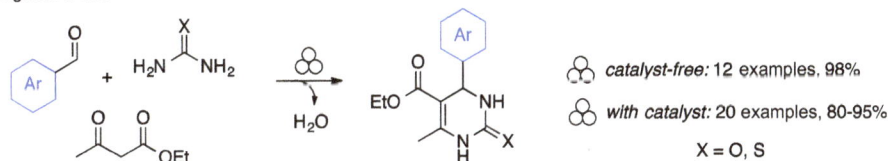

catalyst-free: 12 examples, 98%

with catalyst: 20 examples, 80–95%

X = O, S

Passerini 3-CR   Ugi 4-CR

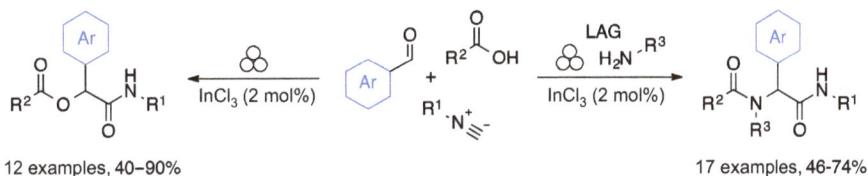

12 examples, 40–90%   17 examples, 46-74%

Strecker 3-CR

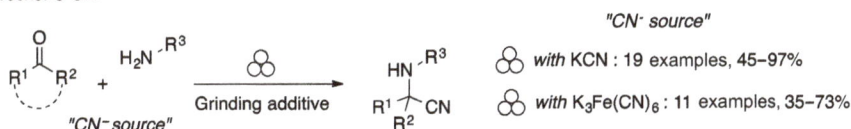

"CN⁻ source"

with KCN : 19 examples, 45–97%

with K₃Fe(CN)₆ : 11 examples, 35–73%

Scheme 4.10: Examples of mechanochemical multicomponent reactions (MCR).

Table 4.10: Process parameters for mechanochemical MCR.

Parameters of the milling process	Method			
	Biginelli[65,a]	**Passerini[67]**	**Ugi[67]**	**Strecker[68, 69,b,c]**
Type of mill	Planetary mill	Mixer mill		Mixer mill
Jar volume (mL)	45	3.2		10,[c] 12,[b] 45[b]
Jar material	SS[d]	Agate, SS[d]	SS[d]	Agate,[b] ZrO₂,[b] W/C [c]
Ball material	SS[d]	Agate, SS[d]	SS[d]	Agate,[b] ZrO₂,[b] W/C[c]
Number of balls	n.r.[e]	2	1	1,[c] 5[b] or 20[b]
Balls diameter (mm)	10	6		5[b,c] or 10[b]
Weight of each ball (g)	47.36[f]	0.44, 1.04[g]	1.04	n.r.[e]
Milling speed	750 rpm	25 Hz		30 Hz
Milling time (min)	30	90		90

Table 4.10 (continued)

Parameters of the milling process	Method			
	Biginelli[65,a]	Passerini[67]	Ugi[67]	Strecker[68, 69,b,c]
Milling/resting period	–	–	–	–
Grinding additive	–	–	MeOH	SiO_2
Reaction scale (mmol)	2		0.5–4	0.5–5

[a]For a *one-pot/two-step* oxidation/Biginelli reaction from alcohols see reference;[66]
[b]with KCN as *CN⁻ source*:[69] agate jar (12 mL) and balls (20 of 5 mm diameter) or ZrO_2 jar (45 mL) and balls (5 of 10 mm diameter);
[c]with $K_3Fe(CN)_6$ as *CN⁻ source*:[68] W/C jar (10 mL) and one ball (10 mm diameter);
[d]SS, stainless steel;
[e]n.r., not reported;
[f]the total weight of the balls is indicated;
[g]for agate (0.44 g), for SS (1.04 g).

4.2.1.5 1,1′-Carbonyldiimidazole-mediated syntheses

1,1′- Carbonyldiimidazole (CDI) is safe and versatile acyl transfer agent easy to handle, compared to the commonly used toxic isocyanates, (tri)phosgene, or chloroformates. Available in kg quantities and cheap, it was successfully applied for the mechanochemical preparation of amides,[71–72] ureas,[73–75] carbammates,[3] hydroxamic acids (methods A–C, Scheme 4.11 and Table 4.11) [76] and hydantoins.[77] It was also a convenient reagent for the mechanochemical preparation of *N*-protected amino ester derivatives[3] (method C) and APIs[72–73, 78] in clean environmental conditions, since it generates relatively innocuous and easy-to-remove by-products (water soluble imidazole and gaseous carbon dioxide).[79]

The bottleneck of the preparation of hydroxamic acids in solution is the low solubility of hydroxylamine hydrochloride in nearly all organic solvent media, resulting in poor reactivity with activated carboxylic acids and the formation of several by-products. As a consequence, the isolation of the water soluble hydroxamic acids is difficult. These drawbacks are overcome by using ball-milling procedures, having the additional advantage of providing higher yields in shorter time and allowing a straightforward recovery of the final products by additional milling of the crude reaction mixture with silica gel (350 mg), followed by filtration (method A).[76]

Similar considerations apply to the CDI-mediated preparation of amides (method B),[72] successfully applied also to the preparation of teriflunomide, FDA-approved drug for treatment of multiple sclerosis.[81] In this case, milling the crude mixture with water (4 mL) led to precipitation of the products, recovered in high yield after filtration.

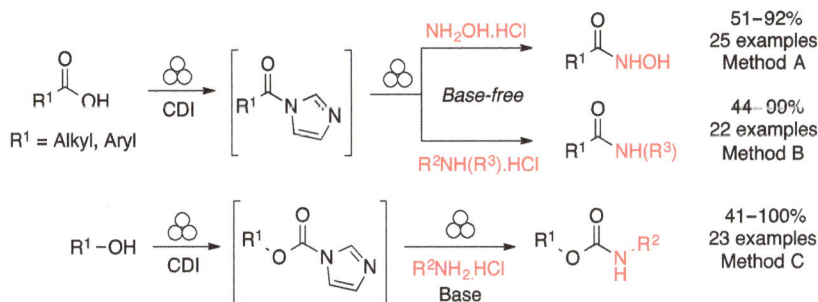

Scheme 4.11: CDI-mediated *one-pot*/two steps mechanochemical synthesis of hydroxamic acids, amides and carbamates.

Table 4.11: Process parameters for mechanochemical CDI-mediated syntheses.

Parameters of the milling process	Method		
	A[76]	B[72]	C1[a]/C2[b,3]
Type of mill	SPEX mixer-mill	Planetary mill	Mixer[a]/planetary[b] mill
Jar volume (mL)	45	12	5[a]/12[b]
Jar material	Agate	SS[c]	SS[c]
Number of balls	4	50	2[a]/25[b]
Balls diameter (mm)	10	5	5
Weight of each ball	n.r.[d]	n.r.[d]	n.r.[d]
Milling speed	18 Hz	500 rpm	30 Hz[a]/450 rpm[b]
Milling time (min)	50[e]	15	105[a]/135[b]
Milling/resting period	–	–	–
Grinding additive	SiO_2[f]	H_2O[f]	Citric acid[f]
Reaction scale (mmol)	1 [g]	1.5 [g,80]	0.6[a]/0.9[b]

[a]Method C1 (mixer-mill): *step 1*: 5 min, *step 2*: 90 min;
[b]Method C2 (planetary ball-mill): *step 1*: 15 min, *step 2*: 120 min;
[c]SS, stainless steel;
[d]n.r., not reported;
[e]*step 1*: 5 min, *step 2*: 45 min;
[f]added during the work-up and milling the crude mixture for additional 3–5 min;
[g]the method is suitable for multi-gram scale preparation.

Mechanochemical procedures were particularly advantageous in the preparation of carbamates, compared to solution-based procedures, usually involving the reaction of the amine in the first step. Due to poor nucleophilicity of alcohols, the second step usually requires (a) the activation of the distal nitrogen of the carbamoyl imidazole by protonation or methylation, (b) the formation of an alkoxide with a strong base (NaOH or NaH 60% in mineral oil) and (c) prolonged heating (Scheme 4.12).[3]

However, preparation of carbamates by mechanochemistry using the "solution-based sequence" was unsuitable. Substrate dependent, variable amount of symmetric ureas were also formed. Therefore, changing the reaction sequence allowed to overcome the limitations of solution-based procedures, additionally providing a general method for carbamate preparation by mechanochemistry. Indeed, no activation of the starting alcohol is needed and the imidazole carboxylic ester is formed directly from the alcohol, activated by mechanical milling. Moreover, the use of the (nucleophilic) amine in the second step also avoids the use of hazardous reactants for activation of the imidazole carboxylic ester intermediate and no urea by-products were formed. Milling the crude mixture with 10% aqueous citric acid led to precipitation of carbamates, filtered to eliminate the nontoxic waste (potassium citrate, used as food additive and imidazole).

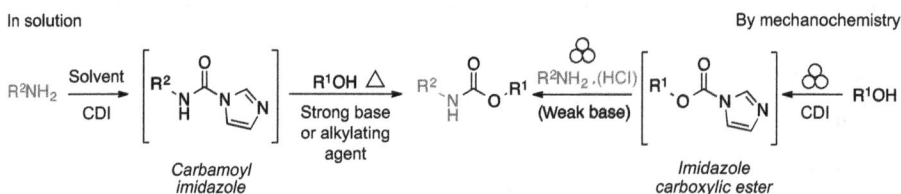

Scheme 4.12: CDI-mediated *one-pot*/two steps mechanochemical synthesis of carbamates: solution vs. mechanochemical procedures.

4.2.1.6 Active Pharmaceutical Ingredients

Medicinal mechanochemistry[82] is an innovative area of investigation opening new horizons for the development of alternative processes with highly reduced environmental impact to access APIs. In this regard, solvent-free mechanochemical procedures present a cheap and sustainable alternative to prepare APIs at a reduced cost, faster and in high yields, with cleaner reaction profiles and simplified work-up procedures, avoiding harmful reaction conditions and complex experimental setup.

The mechanochemical preparation of API perfectly matches with the design of a benign chemistry achieving *economy* of reagents, solvents, time and waste. This is herein illustrated via selected examples, covering transformations not highlighted in the previous sections and possibly completed within a 4 h laboratory period. The aim is to stimulate a sustainable thinking also in the area of drug development and drug discovery. The first mechanochemical synthesis of an API (a metallo-drug) was reported by Friščić and coworkers in 2011: commercialized as antiacid medication under the trade name Pepto-Bismol, the bismuth subsalicylate was obtained by milling salicylic acid with Bi_2O_3.[83]

Pharmaceutically relevant sulfonyl ureas were prepared by copper-catalyzed addition of sulfonyl amide to isocyanates, enabling a simple and fast access to first (tolbutamide and chloropropamide) and second-generation antidiabetic drugs (glibenclamide) with 100% atom economy (Scheme 4.13).[84,85]

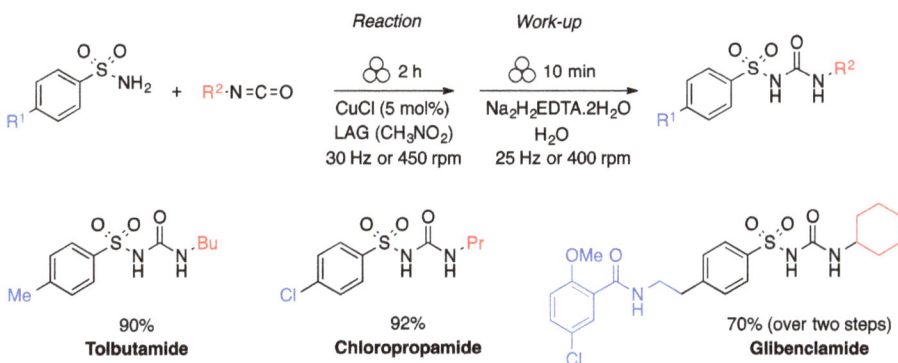

Scheme 4.13: Mechanosynthesis of antidiabetic drugs.

The industrial procedure in solution is a two-step process: a base (NaOH or K_2CO_3 in stoichiometric amount or in excess) activates the sulfonylamide by deprotonation, then its addition to the isocyanate occurs. The mechanochemical procedure is catalytic (*waste economy*), is one step shorter (*step economy*) and is achieved with *reagent economy* and characterized by fast kinetics in LAG using nitromethane. In the case of tolbutamide, the influence of different mechanical stress was investigated using a mixer mill and a planetary mill, both delivering comparable results.[85] Interestingly, two different polymorphic forms could be obtained by milling the reactants in neat or LAG conditions. Worth of note is the straightforward separation of the water-insoluble final products from the copper catalyst. Indeed, milling the crude mixture with Na_2H_2EDTA and water during work-up, generated copper water-soluble salts separated upon filtration.

The mechanochemical preparation of marketed drugs such as nitrofurantoin (antibacterial agent used for the treatment of genitourinary infections) [86] and dantrolene (used for the treatment of malignant hyperthermia),[87] both containing *N*-acyl hydrazone moiety, a functional group extensively used in medicinal chemistry, was also reported.[88] 1-Aminohydantoin hydrochloride was fully condensed with 5-nitro-2-furaldehyde and 5-(4-nitrophenyl)furfural, respectively, affording selectively the *E*-hydrazones in high yields with no need of work up (*waste economy*). Indeed, the pure final powders were recovered by scratching them out directly from the jar, even on the gram scale (Scheme 4.14).[88]

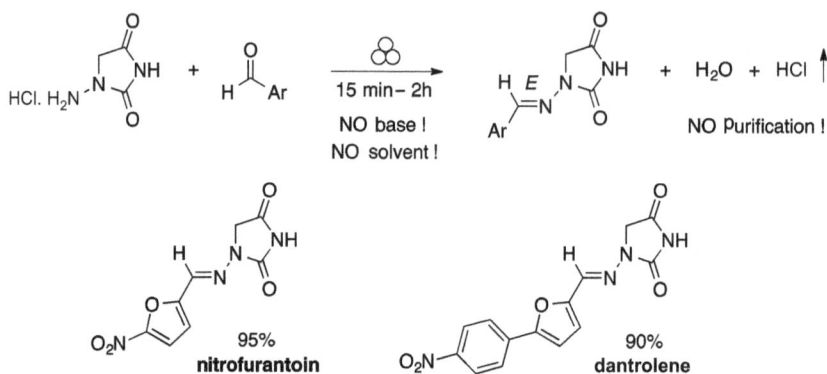

Scheme 4.14: Hydantoin-based *N*-acylhydrazones: preparation of nitrofurantoin and dantrolene.

The addition of a sacrificial base to generate the nucleophilic amine, usually needed in solution to promote the condensation reaction, was not needed[3, 74, 88] (*reagent economy*). The strong activation provided by mechanochemistry is possibly responsible of a proton exchange within the reactants, allowing the reaction to proceed. Moreover, the presence of water and gaseous HCl formed during the synthesis are not able to hydrolyze the new acid-labile hydrazone bond. This is in agreement with previous findings in the literature,[89] suggesting that the water formed during the condensation reaction is present as "crystallization" water. The outcome of the reaction was independent on the milling device used (vibrating, SPEX or planetary mill) and displayed highly improved green metrics and lower synthetic costs, compared to solution based procedures. Moreover, the mechanochemical procedure (a) avoided the use of harsh conditions such as the use of highly concentrated acidic and basic solutions during the reaction and for the work-up, which could be responsible for corrosion problems and raised safety concerns; (b) displayed higher throughput, due to faster kinetics (30 min reaction instead of 8 h in solution); and (c) reduced the energetic costs, the heating or controlled cooling of the mixture being avoided.

An API for the treatment of tuberculosis (Ftivazide) and containing *N*-acyl-hydrazone bond was also prepared by mechanochemistry.[90, 91]

For interested readers, other examples of mechanochemical preparation of API molecules such as axitinib,[19] ethotoin,[73, 78] leu-enkephaline,[92] paracetamol,[93] Phenytoin,[74] procainamide[93] and teriflunomide[72] were also reported.

Conclusions

In comparison with classical thermal methods in solution, a *"benign by design"* chemistry is possible not only by replacing toxic and harmful reagents with safer alternatives, but also applying technological approaches for the sustainable design of the whole process, also avoiding complex experimental set-up. In this regard, mechanochemistry is particularly appealing as it usually provides higher yields, faster kinetics, cleaner reaction profiles (by avoiding the formation of by-products usually encountered in solution) and simplified procedures, and it usually delivers full conversion of stoichiometric amounts of reactants.

Is solvent-based chemistry a religion? No, but the current mindset of chemists and chemical educators needs to change to encompass alternatives to solution-based reactions. In this context, mechanochemistry offers new perspectives for sustainable thinking and conception of a chemical reaction, opening new frontiers for safer and cost-effective manufacturing processes.

References

[1]   Since 2007, Beyond Benign, a non-profit organisation in the USA, co-founded by Dr. John Warner and Dr. Amy Cannon is empowering educators, students and the community to practice sustainability through green chemistry, providing tools, training and support to make green chemistry an integral part of chemistry education since the elementary school. To know more, visit: http://www.beyondbenign.org/, accessed December 2019.
[2]   Anastas, P. T., Farris, C. A., Benign by Design: Alternative Synthetic Design for Pollution Prevention;. ACS Symposium Series 557, American Chemical Society, Washington DC, 1994.
[3]   Lanzillotto, M., Konnert, L., Lamaty, F., Martinez, J., Colacino, E. Mechanochemical 1,1'-carbonyldiimidazole-mediated synthesis of carbamates. ACS Sustainable Chem. Eng. 2015, 3, 2882–2889.
[4]   Ball Milling Towards Green Synthesis: Applications, Projects, Challenges, Stolle, A., Ranu, B., Ed. RSC Green Chemistry Series, 2015.
[5]   Mechanochemistry: From Functional Solids to Single Molecule, Faraday Discuss. RSC, Cambridge, UK, 2014, Vol. 170.
[6]   Margetic, D., Strukil, V. Mechanochemical Organic Synthesis. Elsevier, Amsterdam, NL, 2016, 386.
[7]   Charnay, C., Porcheddu, A., Delogu, F., Colacino, E. New and up-and-coming perspectives for an unconventional chemistry: from molecular synthesis to hybrid materials by mechanochemistry. In: Green Synthetic Processes and Procedures, Edited by Ballini, R., Ed. RSC Green Chemistry Series, 2019, Ch. 9.
[8]   Porcheddu, A., Charnay, C., Delogu, F., Colacino, E. From solution-based non-conventional activation methods to mechanochemical procedures: the hydantoin case. In: 'Non Traditional Activation Methods in Green and Sustainable Applications', Edited by Torok, B., Schafer, C., in 'Advances in Green and Sustainable Chemistry series' (series Editors Bela Torok and Timothy Dransfield), Ed. Elsevier, 2020, in print.
[9]   Rightmire, N. R., Hanusa, T. P. Advances in organometallic synthesis with mechanochemical methods. Dalton Trans. 2016, 45, 2352–2362.
[10]   In this case, the sustainability of a solvent-free mechanochemical process is diminished, due to the waste generated by the large amount of solvents required for the purification.
[11]   Processing waxy or sticky reaction mixtures by ball-milling can be unproductive. A way to circumvent this problem is to make the mixture sufficiently powderly by the addition of grinding auxiliaries (e.g. NaCl, SiO_2, etc.).

[12] For alternative types of set-ups for ball-milling see the supporting information associated to: Colacino, E., Dayaker, G., Morère, A., Friščić, T. Introducing Students to Mechanochemistry via Environmentally Friendly Organic Synthesis Using a Solvent-Free Mechanochemical Preparation of the Antidiabetic Drug Tolbutamide. J. Chem. Educ. 2019, 96, 766–771.

[13] Nielsen, S. F., Peters, D., Axelsson, O. The Suzuki reaction under solvent-free conditions. Synth. Commun. 2000, 30, 3501–3509.

[14] Klingensmith, L. M., Leadbeater, N. E. Ligand-free palladium catalysis of aryl coupling reactions facilitated by grinding. Tetrahedron Lett. 2003, 44, 765–768.

[15] Schneider, F., Ondruschka, B. Mechanochemical solid-state Suzuki reactions using an in situ generated base. ChemSusChem 2008, 1, 622–625.

[16] Schneider, F., Stolle, A., Ondruschka, B., Hopf, H. The Suzuki-Miyaura reaction under mechanochemical conditions. Org. Proc. Res. Devel. 2009, 13, 44–48.

[17] Seo, T., Ishiyama, T., Kubota, K., Ito, H. Solid-state Suzuki–Miyaura cross-coupling reactions: olefin-accelerated C–C coupling using mechanochemistry. Chem. Sci. 2019, 10, 8202–8210.

[18] Tullberg, E., Peters, D., Frejd, T. The Heck reaction under ball-milling conditions. J. Organomet. Chem. 2004, 689, 3778–3781.

[19] Yu, J., Hong, Z., Yang, X., Jiang, Y., Jiang, Z., Su, W. Bromide-assisted chemoselective Heck reaction of 3-bromoindazoles under high-speed ball-milling conditions: synthesis of axitinib. Beilstein J. Org. Chem. 2018, 14, 786–795.

[20] Declerck, V., Colacino, E., Bantreil, X., Martinez, J., Lamaty, F. Poly(ethylene glycol) as reaction medium for mild Mizoroki–Heck reaction in a ball-mill. Chem. Commun. 2012, 48, 11778–11780.

[21] Zhu, X., Liu, J., Chen, T., Su, W. Mechanically activated synthesis of (E)-stilbene derivatives by high-speed ball milling. Appl. Organometal. Chem. 2012, 26, 145–147.

[22] Hermann, G. N., Becker, P., Bolm, C. Mechanochemical rhodium(III)-catalyzed C-H bond functionalization of acetanilides under solventless conditions in a ball Mill. Angew. Chem. Int. Ed. 2015, 54, 7414–7417.

[23] Yu, J., Shou, H., Yu, W., Chen, H., Su, W. Mechanochemical oxidative Heck coupling of activated and unactivated alkenes: A chemo-, regio- and stereo-controlled synthesis of alkenylbenzenes. Adv. Synth. Catal. 2019, 361, 5133–5139.

[24] Colacino, E., Martinez, J., Lamaty, F., Patrikeevaa, L. S., Khemchyan, L. L., Ananikov, V. P., Beletskaya, I. P. PEG as an alternative reaction medium in metal-mediated transformations. Coord. Chem. Rev. 2012, 256, 2893–2920.

[25] Do, J.-L., Mottillo, C., Tan, D., Strukil, V., Friščić, T. Mechanochemical Ruthenium-Catalyzed Olefin Metathesis. J. Am. Chem. Soc. 2015, 137, 2476–2479.

[26] Denlinger, K. L., Ortiz-Trankina, L., Carr, P., Benson, K., Waddell, D. C., Mack, J. Liquid-assisted grinding and ion pairing regulates percentage conversion and diastereoselectivity of the Wittig reaction under mechanochemical conditions. Beilstein J. Org. Chem. 2018, 14, 688–696.

[27] Balema, V. P., Wiench, J. W., Pruski, M., Pecharsky, V. K. Solvent-free mechanochemical synthesis of phosphonium salts. Chem. Commun. 2002, 724–725.

[28] Shearouse, W. C., Mack, J. Diastereoselective liquid assisted grinding: 'cracking' functional resins to advance chromatography-free synthesis. Green Chem. 2012, 14, 2771–2775.

[29] Balema, V., Wiench, J., Pruski, M., Pecharsky, V. Mechanically induced solid-state seneration of phosphorus ylides and the solvent-free Wittig reaction. J. Am. Chem. Soc. 2002, 124, 6244–6245.

[30] Fulmer, D. A., Shearouse, W. C., Medonza, S. T., Mack, J. Solvent-free Sonogashira coupling reaction via high speed ball milling. Green Chem. 2009, 11, 1821–1825.

[31] Thorwirth, R., Stolle, A., Ondruschka, B. Fast copper-, ligand- and solvent-free Sonogashira coupling in a ball mill. Green Chem. 2010, 12, 985–991.
[32] Stolle, A., Ondruschka, B. Solvent-free reactions of alkynes in ball mills: It is definitely more than mixing. Pure Appl. Chem. 2011, 83, 1343–1349.
[33] Trost, B. M. Atom economy. a challenge for organic synthesis. Angew. Chem. Int. Ed. Engl. 1995, 34, 259–281.
[34] Cook, T. L., Walker, J. J. A., Mack, J. Scratching the catalytic surface of mechanochemistry: a multi-component CuAAC reaction using a copper reaction vial. Green Chem. 2013, 15, 617–619.
[35] Mukherjee, N., Ahammed, S., Bhadraa, S., Ranu, B. C. Solvent-free one-pot synthesis of 1,2,3-triazole derivatives by the 'Click' reaction of alkyl halides or aryl boronic acids, sodium azide and terminal alkynes over a Cu/Al$_2$O$_3$ surface under ball-milling. Green Chem. 2013, 15, 389–397.
[36] Rinaldi, L., Martina, K., Baricco, F., Rotolo, L., Cravotto, G. Solvent-free copper-catalyzed azide-alkyne cycloaddition under mechanochemical activation. Molecules 2015, 20, 2837–2849.
[37] Thorwith, R., Stolle, A., Ondruschka, B., Wild, A., Schubert, U. S. Fast, ligand- and solvent-free copper-catalyzed click reactions in a ball mill. Chem. Commun. 2011, 47, 4370–4372.
[38] Tyagi, M., Taxak, N., Bharatam, P. V., Nandanwar, H., Kartha, K. P. R. Mechanochemical click reaction as a tool for making carbohydrate-based triazole-linked self-assembling materials (CTSAMs). Carbohydr. Res. 2015, 407, 137–147.
[39] King, A. O., Okukado, N., Negishi, E. Highly general stereo-, regio-, and chemo-selective synthesis of terminal and internal conjugated enynes by the Pd-catalysed reaction of alkynylzinc reagents with alkenyl halides. J. Am. Chem. Soc. 1977, 19, 683–684.
[40] Colacino, E., Martinez, J., Lamaty, F. The Negishi cross-coupling in the synthesis of natural products and bioactive molecules. In: Modern Tools for the Synthesis of Complex Bioactive Molecules, Cossy, J., Arseniyadis, S., Eds., Johny Wiley and Sons, Hoboken-NJ, USA, 2012, Chap. 2, ISBN: 978-0-470-61618-5.
[41] Cao, Q., Howard, J. L., Wheatley, E., Browne, D. L. Mechanochemical Activation of Zinc and Application to Negishi Cross-Coupling. Angew. Chem. Int. Ed. 2018, 57, 11339–11343.
[42] Grubbs, R. H. Handbook of Metathesis. Wiley-VCH, Weinheim, 2003.
[43] Mukherjee, N., Marczyk, A., Szczepaniak, G., Sytniczuk, A., Kajetanowicz, A., Grela, K. A gentler touch: synthesis of modern ruthenium olefin metathesis catalysts sustained by mechanical force. ChemCatChem 2019, 11, 5362–5369.
[44] Baron, A., Martinez, J., Lamaty, F. Solvent-free synthesis of unsaturated amino esters in a ball-mill. Tetrahedron Lett. 2010, 51, 6246–6249.
[45] Zhang, Z., Peng, Z.-W., Hao, M. F., Gao, J. G. Mechanochemical Diels–Alder cycloaddition reactions for straightforward synthesis of endo-norbornene derivatives. Synlett 2010, 19, 2895–2898.
[46] Tan, Y.-J., Zhang, Z., Wang, F.-J., Wu, -H.-H., Li, Q.-H. Mechanochemical milling promoted solvent-free imino Diels–Alder reaction catalyzed by FeCl$_3$: diastereoselective synthesis of cis-2,4-diphenyl-1,2,3,4-tetrahydroquinolines. RSC Adv. 2014, 4, 35635–35638.
[47] Watanabe, H., Senna, M. Acceleration of solid state Diels–Alder reactions by incorporating the reactants into crystalline charge transfer complexes. Tetrahedron Lett. 2005, 46, 6815–6818.
[48] Watanabe, H., Hiraoka, R., Senna, M. Diels–Alder reaction catalyzed by eutectic complexes autogenously formed from solid state phenols and quinones. Tetrahedron Lett. 2006, 47, 4481–4484.

[49] McKissic, K. S., Caruso, J. T., Blair, R. G., Mack, J. Comparison of shaking versus baking: further understanding the energetics of a mechanochemical reaction. Green Chem. 2014, 16, 1628–1632.

[50] Trotzki, R., Hoffmann, M. M., Ondruschka, B. Studies on the solvent-free and waste-free Knoevenagel condensation. Green Chem. 2008, 10, 767–772.

[51] Trotzki, R., Hoffmann, M. M., Ondruschka, B. The Knoevenagel condensation at room temperature. Green Chem. 2008, 10, 873–878.

[52] Kaupp, G., Naimi-Jamal, M. R., Schmeyers, J. Solvent-free Knoevenagel condensations and Michael additions in the solid state and in the melt with quantitative yield. Tetrahedron 2003, 59, 3753–3760.

[53] Wada, S., Suzuki, H. Calcite and fluorite as catalyst for the Knöevenagel condensation of malononitrile and methyl cyanoacetate under solvent-free conditions. Tetrahedron Lett. 2003, 44, 399–401.

[54] Kulla, H., Haferkamp, S., Akhmetova, I., Roellig, M., Maierhofer, C., Rademann, K., Emmerling, F. In situ investigations of mechanochemical one-pot syntheses. Angew. Chem. Int. Ed. 2018, 57, 5930–5933.

[55] Lukin, S., Tireli, M., Lončarić, I., Barišić, D., Šket, P., Vrsaljko, D., Di Michiel, M., Plavec, J., Užarević, K., Halasz, I. Mechanochemical carbon–carbon bond formation that proceeds via a cocrystal intermediate. Chem. Commun 2018, 54, 13216–13219.

[56] Carta, M., James, S. L., Delogu, F. Phenomenological inferences on the kinetics of a mechanically activated Knoevenagel condensation: understanding the "Snowball" kinetic effect in ball milling. Molecules 2019, 24, 3600–3613.

[57] Burmeister, C. F., Stolle, A., Schmidt, R., Jacob, K., Breitung-Faes, S., Kwade, A. Experimental and computational investigation of Knoevenagel condensation in planetary ball mill. Chem. Eng. Technol. 2014, 37, 857–864.

[58] Schmidt, R., Burmeister, C. F., Baláž, M., Kwade, A., Stolle, A. Effect of reaction parameters on the synthesis of 5-arylidene barbituric acid derivatives in ball mills. Org. Process Res. Dev. 2015, 19, 427–436.

[59] Stolle, A., Schmidt, R., Jacob, K. Scale-up of organic reactions in ball mills: Process intensification with regard to energy efficiency and economy of scale. Faraday Discuss. 2014, 170, 267–286.

[60] Leonardi, M., Villacampa, M., Menéndez, J. C. Multicomponent mechanochemical synthesis. Chem. Sci. 2018, 9, 2042–2064.

[61] Kumar, S., Sharma, P., Kappor, K. K., Hundal, M. S. An efficient, catalyst- and solvent-free, four-component, and one-pot synthesis of polyhydroquinolines on grinding. Tetrahedron 2008, 64, 536–542.

[62] Sachdeva, H., Saroj, R., Khaturia, S., Singh, H. R. Comparative studies of Lewis acidity of alkyl-tin chlorides in multicomponent Biginelli condensation using grindstone chemistry technique. J. Chil. Chem. Soc. 2012, 1, 1012–1016.

[63] Bose, A. K., Pednekar, S., Ganguly, S. N., Chakrabortyb, G., Manhasa, M. S. A simplified green chemistry approach to the Biginelli reaction using grindstone chemistry. Tetrahedron Lett. 2004, 45, 8351–8353.

[64] Tilak, R., Hemant, S., Mayank, A. S., Thammarat, A., Navneet, K., Narinder, S., Doo Ok, J. "Solvent-less" mechanochemical approach to the synthesis of pyrimidine derivatives. ACS Sust. Chem. Eng. 2017, 5, 1468–1475.

[65] M'hamed, M. O., Alshammari, A. G., Lemine, O. M. Green high-yielding one-pot approach to Biginelli reaction under catalyst-free and solvent-free ball milling conditions. Appl. Sci. 2016, 6, 431–436.

[66] Sahoo, P. K., Bose, A., Mal, P. Solvent-Free ball-milling Biginelli reaction by subcomponent synthesis. Eur. J. Org. Chem. 2015, 6994–6998.

[67] Polindara-Garcia, L. A., Juaristi, E. Synthesis of Ugi 4-CR and Passerini 3-CR adducts under mechanochemical activation. Eur. J. Org. Chem. 2016, 1095–1102.

[68] Bolm, C., Mocci, R., Schumacher, C., Turberg, M., Puccetti, F., Hernandez, J. Mechanochemical activation of iron cyano complexes: a Prebiotic impact scenario for the synthesis of a-amino acid derivatives. Angew. Chem. Int. Ed. 2018, 57, 2423–2426.

[69] Hernandez, J., Turberg, M., Schiffers, I., Bolm, C. Mechanochemical Strecker reaction: access to α-aminonitriles and tetrahydroisoquinolines under ball-milling conditions. Chem. Eur. J. 2016, 22, 14513–14517.

[70] Mocci, R., Murgia, S., De Luca, L., Colacino, E., Delogu, F., Porcheddu, A. Ball-milling and cheap reagents breathe green life into the one hundred-year-old Hofmann reaction. Org. Chem. Front. 2018, 5, 531–538.

[71] Kumar, V., Giri, S. K., Venugopalan, P., Kartha, K. P. R. Synthesis of cross-linked glycopeptides and ureas by a mechanochemical, Solvent-Free reaction and determination of their structural properties by TEM and X-ray crystallography. ChemPlusChem 2014, 79, 1605–1613.

[72] Métro, T.-X., Bonnamour, J., Reidon, T., Sarpoulet, J., Martinez, J., Lamaty, F. Mechanosynthesis of amides in the total absence of organic solvent from reaction to product recovery. Chem. Commun. 2012, 48, 11781–11783.

[73] Konnert, L., Dimassi, M., Gonnet, L., Lamaty, F., Martinez, J., Colacino, E. Poly(ethylene) glycols and mechanochemistry for the preparation of bioactive 3,5-disubstituted hydantoins. RSC Adv. 2016, 6, 36978–36986.

[74] Konnert, L., Gonnet, L., Halasz, I., Suppo, J.-S., de Figueiredo, R. M., Campagne, J.-M., Lamaty, F., Martinez, J., Colacino, E. Mechanochemical preparation of 3,5-disubstituted hydantoins from dipeptides and unsymmetrical ureas of amino acid derivatives. J. Org. Chem. 2016, 81, 9802–9809.

[75] Mascitti, A., Lupacchini, M., Guerra, R., Taydakov, I., Tonucci, L., d'Alessandro, N., Lamaty, F., Martinez, J., Colacino, E. Poly(ethylene glycol)s as grinding additives in the mechanochemical preparation of highly functionalized 3,5-disubstituted hydantoins. Beilstein J. Org. Chem. 2017, 13, 19–25.

[76] Mocci, R., De Luca, L., Delogu, F., Porcheddu, A. An environmentally sustainable mechanochemical route to hydroxamic acid derivatives. Adv. Synth. Catal. 2016, 358, 3135–3144.

[77] Colacino, E., Porcheddu, A., Charnay, C., Delogu, F. From enabling technologies to medicinal mechanochemistry: an eco-friendly access to hydantoin-based active pharmaceutical ingredients. Reac. Chem. Eng. 2019, 4, 1179–1188.

[78] Porcheddu, A., Delogu, F., De Luca, L., Colacino, E. From Lossen Transposition to Solventless "Medicinal Mechanochemistry". ACS Sustainable Chem. Eng. 2019, 7, 12044–12051.

[79] Partially readapted from references 3 and 78.

[80] Métro, T.-X., Bonnamour, J., Reidon, T., Duprez, A., Sarpoulet, J., Martinez, J., Lamaty, F. Comprehensive study of the organic-solvent-free CDI-mediated acylation of various nucleophiles by mechanochemistry. Chem. Eur. J. 2015, 21, 12787–12796.

[81] O'Connor, P., Wolinsky, J. S., Confavreux, C., Comi, G., Kappos, L., Olsson, T. P., Benzerdjeb, H., Truffinet, P., Wang, L., Miller, A., Freedman, M. S. Randomized trial of oral Teriflunomide for relapsing multiple sclerosis. New Engl. J. Med. 2011, 365, 1293–1303.

[82] Tan, D., Loots, L., Friščić, T. Towards medicinal mechanochemistry: evolution of milling from pharmaceutical solid form screening to the synthesis of active pharmaceutical ingredients (APIs). Chem. Commun. 2016, 52, 7760–7781.

[83] André, V., Hardeman, A., Halasz, I., Stein, R. S., Jackson, G. J., Reid, D. J., Duer, M. J., Curfs, C., Duarte, M. T., Friščić, T. Mechanosynthesis of the Metallodrug Bismuth Subsalicylate from Bi_2O_3 and Structure of Bismuth Salicylate without Auxiliary Organic Ligands. Angew. Chem. Int. Ed. Engl. 2011, 50, 7858–7861.

[84] Tan, D., Strukil, V., Mottillo, C., Friščić, T. Mechanosynthesis of pharmaceutically relevant sulfonyl-(thio)ureas. Chem. Commun. 2014, 50, 5248–5250.

[85] Colacino, E., Dayaker, G., Morère, A., Friščić, T. Introducing Students to Mechanochemistry via Environmentally Friendly Organic Synthesis Using a Solvent-Free Mechanochemical Preparation of the Antidiabetic Drug Tolbutamide. J. Chem. Educ. 2019, 96, 766–771.

[86] Guay, D. R. An update on the role of nitrofurans in the management of urinary tract infections. Drugs 2001, 61, 353–364.

[87] Krause, T., Gerbershagen, M. U., Fiege, M., Weisshorn, R., Wappler, F. Dantrolene–a review of its pharmacology, therapeutic use and new developments. Anaesthesia 2004, 59, 364–373.

[88] Colacino, E., Porcheddu, A., Halasz, I., Charnay, C., Delogu, F., Guerra, R., Fullenwarth, J. Mechanochemistry for "no solvent, no base" preparation of hydantoin-based active pharmaceutical ingredients: nitrofurantoin and dantrolene. Green Chem. 2018, 20, 2973–2977.

[89] Kaupp, G. Organic Solid-State Reactions with 100% Yield. Top. Curr. Chem. 2005, 254, 95–183.

[90] Oliveira, P. F. M., Baron, M., Chamayou, A., André-Barrès, C., Guidetti, B., Baltas, M. Solvent-free mechanochemical route for green synthesis of pharmaceutically attractive phenol-hydrazones. RSC Adv. 2014, 4, 56736–56742.

[91] Oliveira, P. F. M., Guidetti, B., Chamayou, A., André-Barrès, C., Madacki, J., Korduláková, J., Mori, G., Orena, B. S., Chiarelli, L. R., Pasca, M. R., Lherbet, C., Carayon, C., Massou, S., Baron, M., Baltas, M. Mechanochemical synthesis and biological evaluation of novel isoniazid derivatives with potent antitubercular activity. Molecules 2017, 22, 1457–1484.

[92] Bonnamour, J., Métro, T.-X., Martinez, J., Lamaty, F. Environmentally benign peptide synthesis using liquid-assisted ball-milling: application to the synthesis of Leu-enkephalin. Green Chem. 2013, 15, 1116–1120.

[93] Portada, T., Margetic, D., Štrukil, V. Mechanochemical catalytic transfer hydrogenation of aromatic nitro derivatives. Molecules 2018, 23, 3163–3180.

Ivan Halasz

5 Mechanochemical preparation of cocrystals

There has been much dispute over the formal definition of cocrystals.[1–5] While we shall not engage here in this discussion, we wish to emphasize certain characteristics common to cocrystals as we will encounter them in this chapter: (i) Cocrystals are multicomponent molecular solids where the molecules are held together by supramolecular interactions, (ii) cocrystals have a defined chemical composition, (iii) molecules comprising cocrystals are in general electrically neutral, or otherwise the compound in question could better be termed as a salt and (iv) the cocrystal components, that is, cocrystal coformers, will most often be solid at ambient conditions and the cocrystal itself will be solid at ambient conditions too.

Having established the general boundaries of the compounds in question, it is clear that this chapter will not be focused on breaking of covalent bonds or on making new ones. We will rather study rearrangement of supramolecular interactions and how this can be achieved by mechanochemical milling as well as provide a perspective on how this can be exploited in the materials science. Supramolecular interactions are the forces that hold together the molecules in a cocrystal (as well as in pure compounds), determine their relative orientations and the way molecules pack in the crystals and thus their crystal structures. These interactions may be of the hydrogen- or halogen-bonding type, which are the strongest intermolecular interactions and most exploited in crystal engineering and the design of molecular solids.[6]

Preparation of cocrystals was among the first areas of application of mechanochemistry where the potential of mechanochemistry was recognized, where milling was extensively used, and wherefrom some novel and advanced concepts of mechanochemistry have emerged.[7–10] Indeed, preparation of cocrystals is an area where mechanochemistry excels.[11] In comparison with solution-based approach to preparation of cocrystals, mechanochemistry is often faster and not restricted by the solubility of the cocrystal coformers or of the target cocrystal. It is thus possible to use mechanochemistry for efficient screening[12, 13] and to prepare cocrystals of poorly soluble components, as is for example, theobromine, a popular cocrystal coformer.[14] Mechanochemistry is also highly beneficial in preparing cocrystals with more than two components. Three-component cocrystals[15, 16] can thus be systematically prepared and explored in the solid state.[17] From solution, in general, the least soluble cocrystal will be isolated and it does not need to correspond to the composition of the solution. For example, if two coformers are capable of forming two cocrystals with compositions of say, 1:1 and 1:2, the least soluble cocrystal will be isolated from the solution, if not always, then from a range of coformer solution concentrations. Such a limitation is usually not at play with mechanochemical preparation of cocrystals and, in general, the composition of the isolated cocrystal will correspond to the composition of the starting mixture, provided of

https://doi.org/10.1515/9783110608335-005

course, that such cocrystals with different stoichiometries of the coformers exist and are stable relative to the pure starting coformer phases.

Cocrystal formation starts with planning the target composition of a cocrystal, preparing a mixture of solid reactants in the target stoichiometric ratio and subjecting it to mechanochemical agitation, here by ball milling. Milling can be conducted neat, that is with no additives or with additives, which are usually liquids.[18] In the case liquids are used, milling is most commonly termed liquid-assisted grinding (LAG),[17] though older terms such as solvent-drop grinding,[19] kneading[20] or simply wet grinding may still be encountered. There is no strict rule on the amount of liquid that is to be added, but it should better not make the reaction mixture wet enough so that it turns into a paste. The amount of the added liquid may empirically be expressed by the so-called η parameter, described in Chapter 2, giving the ratio of the added liquid volume (in µL) and the total mass of the reaction mixture (in mg).[10] For a LAG experiment, the η value will be usually be in the range 0.1–0.5 µL/mg. The liquid additive often provides a beneficial catalytic effect accelerating the reaction kinetics and sometimes templating specific product formation. However, little is still known on the mechanism of its catalytic action.

Common practice is to mill the mixture of coformers for 30 min or 1 h after which the reaction mixture will be analyzed by, for example, powder X-ray diffraction (PXRD) or infrared spectroscopy. Characterization by techniques that require dissolution of the cocrystal will be less informative because cocrystals give mixtures of components upon dissolution. With the advances in methodology for crystal structure solution from powder diffraction data, it is often not necessary to prepare single crystals of the cocrystals for their structural characterization. While the precision of crystal structures will in general be better from single-crystal diffraction data, powder diffraction will usually provide crystal structures of sufficient quality to enable the analysis of supramolecular interactions and packing. Approach to crystal structure solution via powder diffraction not only circumvents the necessity to prepare single crystals, which may be directed by seeding,[21] but also enables analysis of the bulk of the product which was not altered in any way after its mechanochemical preparation.

5.1 Hydrogen-bonded cocrystals

We start by describing a model system of hydrogen-bonded cocrystal formation between two simple molecules, benzoic acid (ba) and nicotinamide (na) (Scheme 5.1).

The target cocrystal between ba and na has the composition 1:1. Its formation can be tested using neat grinding (NG) or LAG with various liquids chosen to cover polar, nonpolar, protic and aprotic liquids (Tables 5.1 and 5.2). The cocrystal is easily formed, usually within 30 min of milling at the 1 mmol scale. This reaction offers an

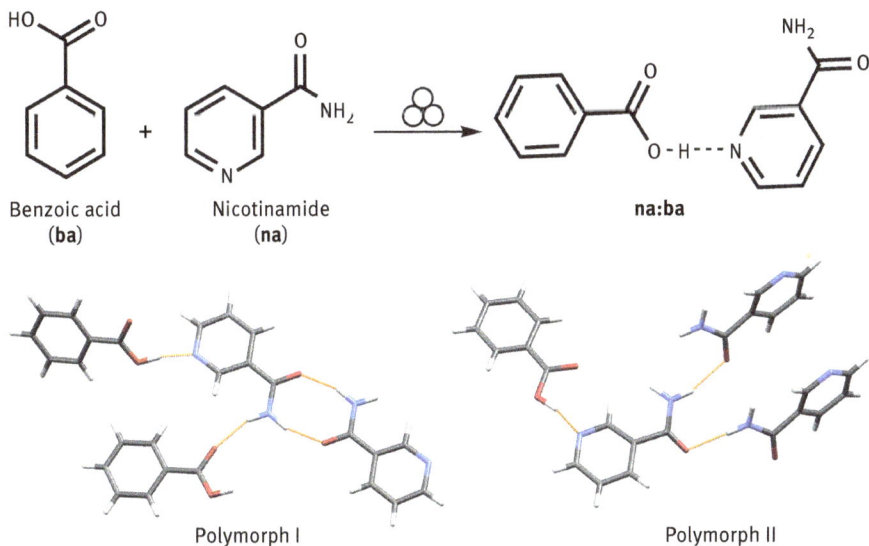

Scheme 5.1: Cocrystal formation between benzoic acid and nicotinamide (top) and fragments of hydrogen-bonded networks in the cocrystals of two polymorphs (bottom). Reproduced from reference[22] with permission.

Table 5.1: Process parameters for mechanochemical cocrystal formation between benzoic acid and nicotinamide.

Key parameters of the milling process	Method	
	Mixer mill	Planetary mill
Type of mill	Mixer mill	Planetary mill
Jar volume (mL)	10–25	50
Jar material	SS[a]	SS[a]
Number of balls	2	12
Balls diameter (mm)	7	10
Weight of each ball (g)	1.4	4.0 g
Milling speed	30 Hz	400 rpm
Milling time (min)	30	30
Grinding additive	None or liquid	None or liquid
Reaction scale (mmol)	1–2	5

[a]SS, stainless steel.

Table 5.2: Selection of liquids for liquid-assisted grinding and their properties.

LIQUID	Polar	Protic	DN[a]	BP[b] (°C)	ε[d]
Methanol	•	•	19.0	64.6	32.7
Ethanol	•	•	19.2	78.5	24.7
1-Propanol		•	19.8	97.0	20.10
2-Propanol		•	21.1	82.4	20.18
Nitromethane	•		2.7	101.2	34.82
Acetonitrile	•		14.1	81.6	37.5
Dioxane			14.8	101.1	2.25
2-Butanone			17.4	79.6	18.5
Ethyl acetate			17.1	77.1	6.02
Water	•	•	18.0	100.0	80.1
Dichloromethane			1.0	39.8	9.1
Nitrobenzene[c]	•		4.4	101.0	34.82
DMF[c]	•		26.6	153.0	36.7
Acetone	•		17.0	56.3	20.7
Triethylamine			61.0	88.6	2.42
Diethylether			19.2	34	4.33
Pyridine			33.1	115.2	12.4

[a]DN, Gutmann's donor number – a measure of basicity;[23]
[b]BP, boiling point;
[c]low-vapor pressure liquid;
[d]εrelative permitivity.

additional advantage as the cocrystal can be formed in three stable polymorphs.[22] If PXRD is available, this is the most convenient method to characterize the sample in terms of the isolated polymorphic form.

Infrared spectroscopy can be used to verify that no proton transfer has occurred by comparing the spectra of pure components to the spectrum of the cocrystal. While there certainly will be changes in the IR spectra, the most significant band of carboxylic group vibrations will remain at approximately the same position. In case of proton transfer and formation of the carboxylate group, this shift would be significant.

5.2 Selectivity in cocrystal formation

Formation of a cocrystal is determined by its stability relative to the starting phases. In an extensive study, Taylor and Day have made a survey of the stability of cocrystals and found that, while there are a few outliers, in general, the cocrystal is more stable than the pure components, that is, cocrystallization is thermodynamically favourable.[24] If a cocrystal between two components cannot be found it will likely be due to the potential cocrystal phase being less stable than pure components.[25] In such cases, there is no experimentally known cocrystal phase that could be used to calculate its energy, along with energies of the starting phases, but the energy of a potential cocrystal must be reached by first predicting its crystal structure.

There are however peculiar examples where cocrystal formation was hindered by nucleation. Bučar et al. have described an elusive cocrystal between caffeine and benzoic acid that was only made after heteronuclear seeds have lowered the barrier for its formation.[26] It is noteworthy that the caffeine:benzoic acid cocrystal was predicted to be more stable than pure components, but despite decades of attempts using various established cocrystal screening methods, including mechanochemical milling, a cocrystal failed to materialize. That is, until it was prepared for the first time. Once prepared, there were no obstacles to reproduce its preparation in the same laboratory and the cocrystal formation could readily be achieved to an extent that it was difficult to maintain a physical mixture of caffeine and benzoic acid. This brings to mind remarkable cases of disappearing polymorphs, some of which also had a commercial and medical impact.[27, 28]

Stability of certain polymorphic forms of a cocrystal during milling can be altered by the use of liquid additives. Belenguer *et al.* have shown that cocrystal nanoparticles may be stabilized in a specific polymorphic form by using a liquid additive,[29] similar to mechanochemical selectivity between polymorphs of molecular compounds[30] (Figure 5.1). It was found that switching the stability between two polymorphs is achieved upon addition of a sufficient amount of a given liquid, this amount being different for different liquids. The observation was explained by stabilization of the surface of milled nanoparticles through interaction with the liquid. By changing the liquid and its amount, surface energy of nanoparticles can be altered and polymorph selectivity switched.

Selectivity in formation of cocrystals will often obey the Ostwald's rule of stages with phases appearing in the order of stability thus determining the sequence of occurrence of cocrystal phases. This is wherefrom our second suggested experiment comes.[32] First, the 1:1 cocrystal of nicotinamide (na) and anthranilic acid (ana) is prepared by milling using a vibratory ball mill at 30 Hz. The as prepared nicotinamide:anthranilic acid cocrystal (naana) is offered 1 equivalent of salicylic acid (sal), another cocrystal coformer. The 1:1 cocrystal of nicotinamide and salicylic acid (nasal) is more stable than the naana cocrystal and sal should therefore push out ana from the naana cocrystal to form the nasal cocrystal and

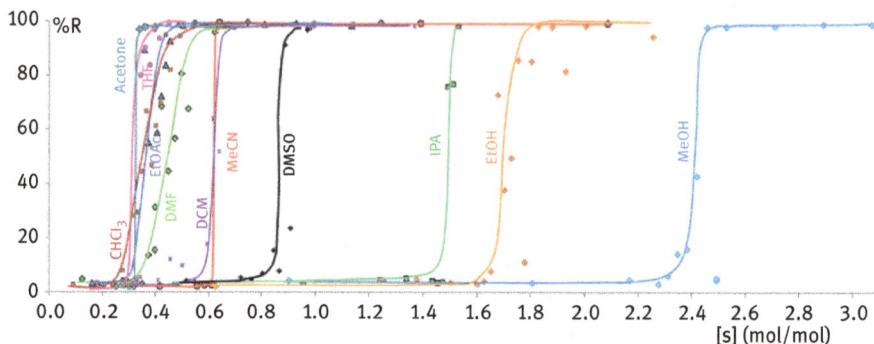

Figure 5.1: Conversion between cocrystal polymorphs using different liquids in a range of amounts. Some liquids sharply change polymorphs while for some the two polymorphs can coexist in a certain range of added liquid volumes.[31] Curves are drawn as a guide to the eye. Reproduced from reference[29] with permission from the Royal Society of Chemistry.

the free ana. This experiment is performed on the 1 mmol scale. At the end of milling after 30 min, the reaction mixture will contain the nasal cocrystal and ana as separate solids.

In case milling is interrupted after 5 min, and the reaction mixture is analyzed fast by PXRD, it should be possible to establish the presence of the intermediate cocrystal between salicylic acid and anthranilic acid (salana) (Figure 5.2). As sal pushes out ana from naana to form nasal, the as formed ana will react with the present sal to form the salana cocrystal. However, as sal becomes fully incorporated in both salana and nasal cocrystals, and with the remaining naana present, the salana disintegrates to react with naana and to finally form the nasal cocrystal and pure ana.

This sequence follows the stability of various possible compositions. Any combination of cocrystal with the third phase present as a pure phase is more stable than the three pure phases. In other words, all cocrystals are more stable than pure components and the most stable is the combination of nasal and pure ana. It is noteworthy that ana can exist in three polymorphs.[33] Usually, polymorph I (ana I) is obtained commercially. As ana is replaced from the naana cocrystal, it crystallized as the ana III polymorph instead of the starting ana I. This is contrary to the expected reactivity based on stabilities of phases, but is in accordance with previous observation by Trask et al. who studied polymorphic transformations of ana and have found that polymorph III is obtained upon grinding of the polymorph I.[30]

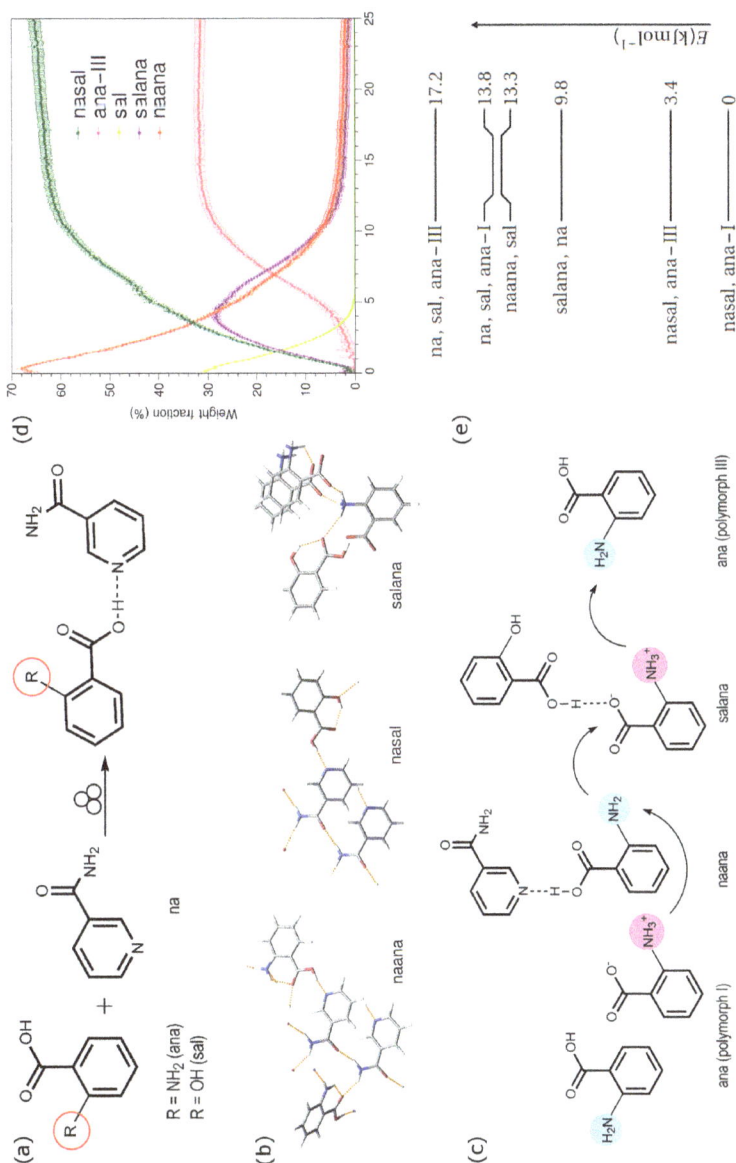

Figure 5.2: (a) Reaction scheme or the formation of the nasal and naana cocrystals. (b) Fragments of crystal structures of the cocrystals with the hydrogen-bonding network denoted with orange dashed lines. The salana cocrystal is zwitterionic with the ana molecule being in the zwitterionic form. (c) Changes of the protonation state of ana as it travels between phases in the studied mechanochemical transformations. Starting commercial form of ana was the polymorph I.[33] (d) Reaction profile of the reaction of salicylic acid with the nicotinamide:anthranilic acid cocrystal, derived from *in situ* reaction monitoring by synchrotron powder X-ray diffraction. (e) Stabilities of various compositions of the reaction mixture comprised of na, sal and ana, including possible cocrystals and corresponding pure components. Energy is given in kJ mol^{-1} and was estimated using solid-state DFT calculations. Anthranilic acid is known to exist in three polymorphic forms, polymorphs I and III appear here. Adapted with permission from reference[32] (Copyright (2018) American Chemical Society).

5.3 Halogen-bonded cocrystals

Another strong interaction is the halogen bond. Equally to hydrogen bonding, halo-gen bonding can be employed in crystal engineering and to prepare novel cocrystals. A recent comparison of hydrogen and halogen bonds has shown that halogen bond can easily be stronger than hydrogen bonds and may thus dominate in some crystal engineering attempts.[34] For the preparation of halogen bonded cocrystals we have here chosen a system capable of providing cocrystals with stoichiometries 1:1 and 1:2 by using a 4,4'-bipyridine, a halogen bond acceptor possessing two acceptor sites, and N-bromosuccinimide as halogen bond donor (nbs, Figure 5.3).[35]

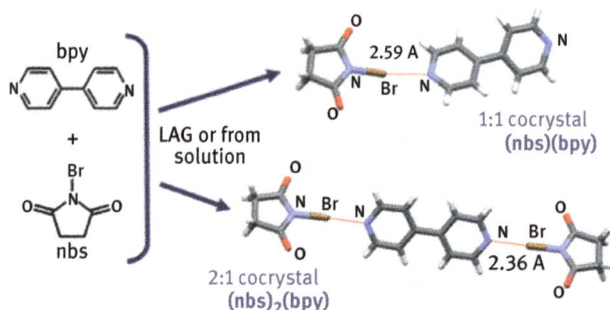

Figure 5.3: Formation of 1:1 and 1:2 halogen-bonded cocrystals using N-bromosuccinimide (nbs) as the halogen bond donor and 4,4'-bipyridine (bpy) as the halogen bond acceptor.[35] Reproduced from reference[35] with permission from the Royal Society of Chemistry.

The stoichiometry of the resulting cocrystal is controlled with the input stoichiomet-ric ratio of solid reactants. In NG, the 1:1 cocrystal is prepared within 15 min milling on a vibratory ball mill operating at 25 Hz and starting from the 1:1 reactant mixture. The 1:2 cocrystal requires significantly longer milling times to be prepared pure under NG. Facile formation of the 1:1 cocrystal occurs also with the 1:2 initial reac-tant ratio and the 1:1 cocrystal can be observed ex situ as an intermediate during preparation of the 1:2 cocrystal. Formation of the 1:2 cocrystal is accelerated by LAG using acetonitrile and the 1:2 cocrystal can also be prepared within 15 min milling.

Table 5.3: Process parameters for mechanochemical formation of halogen-bonded cocrystals between *N*-bromosuccinimide and 4,4′-bipyridine.

Key parameters of the milling process	Method	
Reactant ratio (bipy:nbs)	1:1	1:2
Type of mill	Mixer-mill	Mixer mill
Jar volume (mL)	10–25	10–25
Jar material	SS[a]	SS[a]
Number of balls	2	2
Balls diameter (mm)	7	7
Weight of each ball (g)	1. 4	1.4
Milling frequency	25	25
Milling time (min)	15	NG: 90 LAG[b]: 15
Grinding additive	None or liquid	None or liquid
Reaction scale (mmol)	1	1

[a]SS, stainless steel;
[b]liquid additive may be acetonitrile.

5.4 Pharmaceutical cocrystals

An important area where cocrystals find application is the pharmaceutical industry,[36–38] since cocrystals of active pharmaceutical ingredient (APIs) were recognized to be equally viable in modifying the API properties as solvate and salt formation previously were. The API may lack certain desirable or required physical properties such as good solubility, bioavailability or stability. For example, solubility and bioavailability of a poorly soluble drug can be improved if the API is crystallized as a salt. However, not all APIs can be crystallized as salts due to lacking appropriate functional groups that could enable an ionic form of the API molecule.

Since recently, cocrystal formation of pharmaceutical compounds has been added to the toolbox of possible modification for the preparation of API formulations with improved properties.[39, 40] Pharmaceutical cocrystals are thus compounds just as "ordinary" cocrystals except for the fact that at least one component needs to be an API. Other components of a pharmaceutical cocrystal should be taken from a group of compounds that are "generally recognized as safe", so that the cocrystal could potentially be used as a medication.[41]

5.4.1 Cocrystal of vitamin C and nicotinamide

As the first example of a pharmaceutical cocrystal we present the 1:1 cocrystal of two vitamins, vitamin C (ascorbic acid, abbreviated as acs) and vitamin B3 (nicotinamide, abbreviated as na). This cocrystal is now known for decades and is approved for human use in multivitamin preparations.[42] Its preparation seems to have been first described in the literature in 1945 where it was prepared from solution, but also by ball milling. Only then, it was not called a cocrystal but rather a molecular addition compound. Formation of the (asc)(na) cocrystal can also be observed visually, since it is yellowish, while the cocrystal components are white. On a vibratory ball mill, cocrystal synthesis can usually be achieved on the 1 mmol scale within 30 min milling and by using ethanol as the liquid additive (Figure 5.4, Table 5.4).[43] If methanol is used, a second polymorph may be observed and when using more methanol, the reaction becomes faster and more selective toward the formation of the second polymorph. If available, reaction vessels made from a transparent material (such as the plastic poly(methyl methacrylate)) could be used to directly observe the color change during milling. This reaction is also suitable for a scale-up experiment since reagents are affordable and the reaction has been tested on the 10 g scale on a planetary ball mill as well as in a continuous process on a twin-screw extruder.

Pharmaceutical cocrystals as well exhibit selectivity. An API that forms cocrystals with two components can be expected to exhibit selectivity. Competition of salicylic acid and anthranilic acid for cocrystal formation with sulfadimidine (one of sulfamide drugs) was the subject of a pioneering study in 1995 by Caira and coworkers who demonstrated that ana replaces sal from the sulfadimidine:sal cocrystal by grinding in the solid state and providing as a result a mixture of the sulfadimidine:ana cocrystal and pure sal[44] (Scheme 5.2 and Table 5.5).

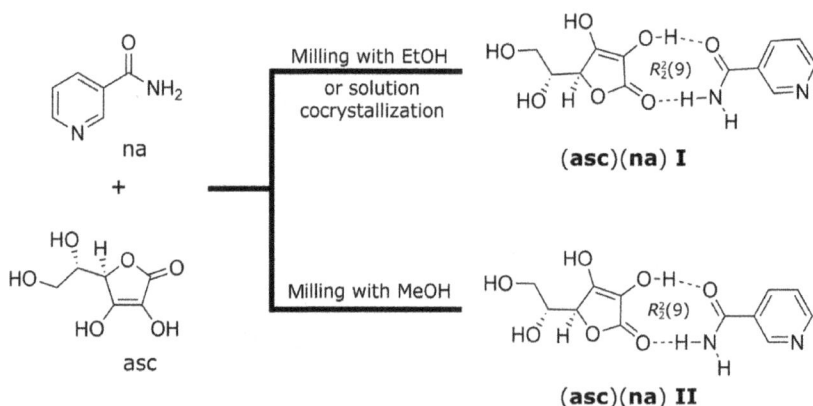

Figure 5.4: Formation of the 1:1 cocrystal of vitamin C (asc) and nicotinamide (na). Using ethanol or methanol as liquids in a LAG reaction, two polymorphs can be prepared (denoted as I or II). A segment of hydrogen bonding network is given along with its graph-set notation. Reprinted with permission from reference[43] (Copyright (2019) American Chemical Society).

Table 5.4: Process parameters for mechanochemical formation of a pharmaceutical cocrystal between vitamin C and vitamin B3.

Parameters of the milling process	Method		
Reactant ratio (asc:na)	1:1	1:1	1:1
Type of mill	Mixer mill	Mixer mill	Planetary
Jar volume (mL)	ca. 14	ca. 14	50
Jar material	SS[a], PMMA[b]	SS[a], PMMA[b]	ZrO_2, SS[a]
Number of balls and type	2, SS	2, SS	4, ZrO_2
Balls diameter (mm)	7	7	10
Weight of each ball (g)	1.4	1.4	3.0
Milling speed	30 Hz	30 Hz	400 rpm
Milling time (min)	10–30	15–40	60
Grinding additive	MeOH	MeOH	EtOH
Liquid volume	20–40 µL	20–40 µL	0.8–1.2 mL
Reaction scale	1 mmol	1 mmol	10 g

[a]SS, stainless steel;
[b]PMMA, poly(methyl methacrylate).

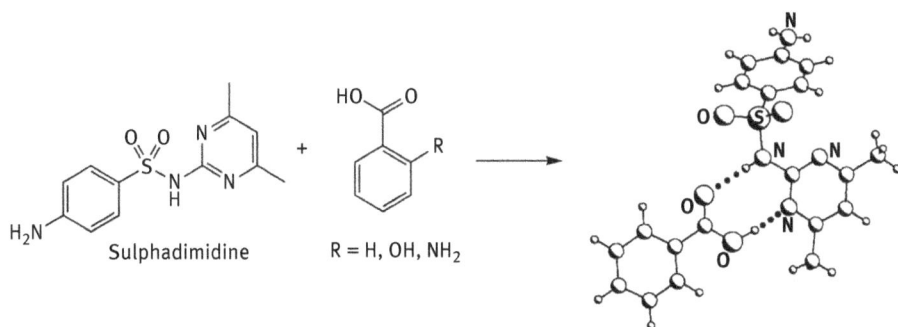

Scheme 5.2: Cocrystal formation between sulfadimidine and benzoic acid and its derivatives. To the right is given a hydrogen-bonding motif in the cocrystals. Adapted from reference[44] (reprinted in part with permission from the Royal Society of Chemistry).

If sulfadimidine is available, this experiment can be performed in a similar manner as previous selectivity experiments with cocrystals of nicotinamide and salicylic acid or anthranilic acid. Sulfadimidine can be reacted with the equimolar mixture or ana and sal to yield the sd:ana cocrystal in the mixture with sal. In an another experiment, the sd:sal cocrystal is first prepared and then offered an equivalent of ana to observe replacement of sal from the cocrystal and formation of the sd:ana cocrystal. The sal:ana cocrystal, previously encountered in the selectivity study between na:ana and na:sal cocrystals, may here as well be observed as an intermediate or a smaller amount of it still residing in the final reaction mixture.

Table 5.5: Process parameters for mechanochemical formation of a pharmaceutical cocrystal between sulfadimidine (sd) and anthranilic acid (ana) or salicylic acid (sal).

Parameters of the milling process	Method	
Reactant ratio (sd:sal:ana)	1:1:1	1:1:0[a]
Type of mill	Mixer mill	Mixer mill
Jar volume (mL)	10–15	10–15
Jar material	SS[b], PMMA[c]	SS[b], PMMA[c]
Number of balls and type	2, SS	2, SS
Balls diameter (mm)	7	7
Weight of each ball (g)	1.4	1.4
Milling speed	30 Hz	30 Hz
Milling time (min)	30	40
Grinding additive	None	None
Reaction scale	0.5 mmol	0.5 mmol

[a]Followed by addition of 1 equivalent (0.5 mmol) of ana after preparing the sd:sal cocrystal;
[b]SS, stainless steel;
[c]PMMA, polymethyl methacrylate.

We can compare directly NG and LAG for the preparation of pharmaceutical cocrystals on the 1:1 cocrystal of carbamazepine (cbz) and saccharine (sac).[45] These reactions were the subject of an in situ monitoring study using PXRD. The (cbz)(sac) cocrystal is readily prepared within minutes of milling when using a liquid additive, for example acetonitrile (Figure 5.5).

In NG however, no reaction is observed, but rather the reaction mixture becomes gradually amorphous.[46] In the case of NG, X-ray diffraction is not appropriate to determine if a chemical reaction is taking place in the amorphous phase, but revealed that the crystallinity of the reaction mixture is being lost upon mechanochemical processing, and that no crystalline products appear. In situ PXRD monitoring of a LAG experiment on the other hand revealed fast product formation and the reaction, on the 1 mmol scale, was complete within several minutes of milling.

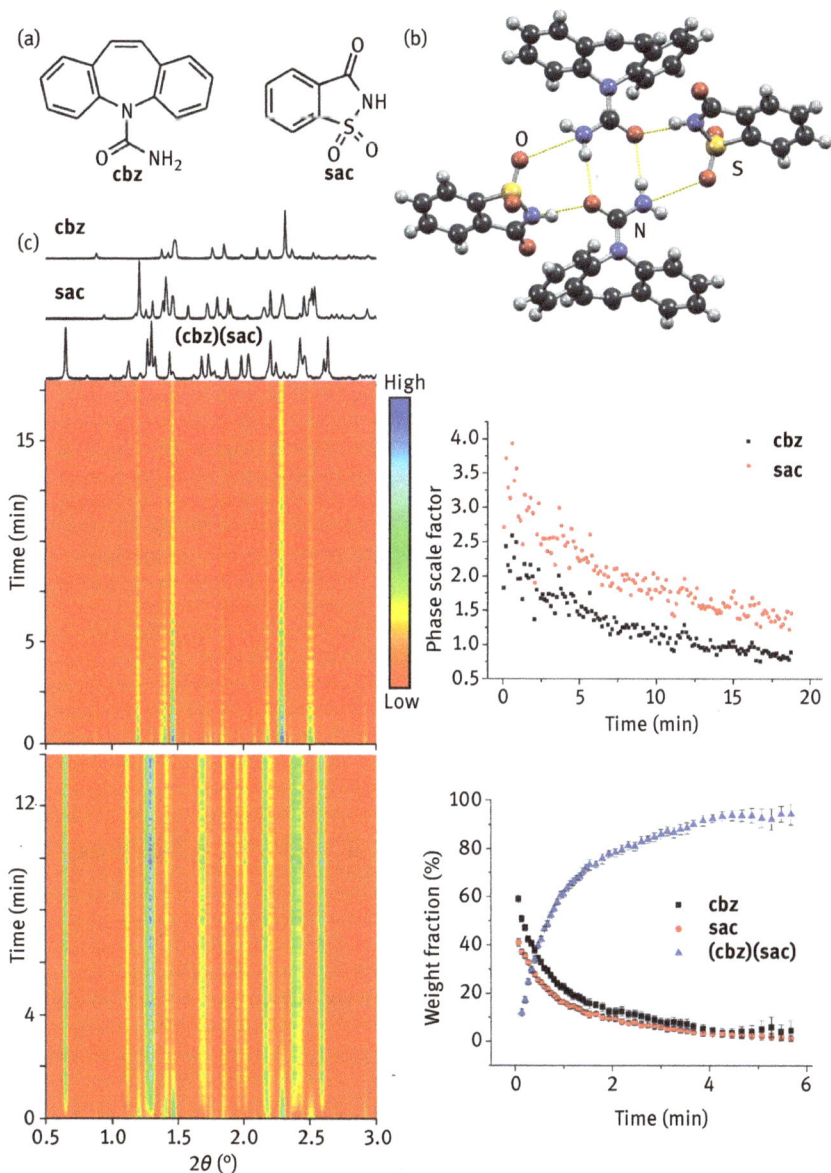

Figure 5.5: (a) molecular structures of carbamazepine and saccharine. (b) Fragment of the crystal structure of their cocrystal with the hydrogen bonding network denoted in yellow thin lines. (c) 2-dimensional time-resolved diffractograms derived from in situ reaction monitoring of (top) a NG reaction between carbamazepine and saccharine which yielded no new crystalline phases and (bottom) LAG using acetonitrile providing the 1:1 (cbz)(sac) cocrystal within minutes of milling. Reaction profiles are given to the right of the 2-dimensional diffractograms. Adapted from reference[46] with permission from Wiley.

In the laboratory, the reaction can be tested on the 1 mmol scale using process parameters described earlier for other reactions. The reaction is robust and variations in reaction conditions and process parameters can be tolerated.

Conclusions

The limited selection of examples in this chapter has intended to demonstrate that mechanochemical milling is becoming a widely popular approach for the preparation of cocrystals since it provides an efficient synthetic route with fast and scalable reactions, clean processes with the target cocrystals normally recovered pure and in quantitative yield, products with well controlled stoichiometry and without limitations imposed by the solubility of the cocrystal or its components, as well as with a significant control over the resulting polymorphic form. Mechanochemical approach strongly benefits from the use of liquid additives which greatly expand its versatility as well as contribute to controlling the crystallinity of the recovered products. With these benefits, and with cocrystals becoming recognized as a means to modification of an API, it can be expected that solvent-free or low-solvent mechanochemical cocrystal preparation will soon find its place in industrial manufacturing.

References

[1] Aakeröy, C. B., Salmon, D. J. Building co-crystals with molecular sense and supramolecular sensibility. Cryst. Eng. Comm. 2005, 7, 439–448.

[2] Bond, A. D. What is a co-crystal? Cryst. Eng. Comm. 2007, 9, 833–834.

[3] Desiraju, G. R. Counterpoint: What's in a Name? Cryst. Growth Des. 2004, 4, 1089–1090.

[4] Dunitz, J. D. Crystal and co-crystal: a second opinion. Cryst. Eng. Comm. 2003, 5, 506.

[5] Thayer, A. M. Dispute Over Crystal Structure Nomenclature Takes Center Stage. Chem. Eng. News 2007, 85, 28–29.

[6] Corpinot, M. K., Bučar, D. K., Practical, A. Guide to the Design of Molecular Crystals. Cryst. Growth Des. 2019, 19, 1426–1453.

[7] Vishweshwar, P., McMahon, J. A., Peterson, M. L., Hickey, M. B., Shattock, T. R., Zaworotko, M. J. Crystal engineering of pharmaceutical co-crystals from polymorphic active pharmaceutical ingredients. Chem. Commun. 2005, 4601–4603.

[8] Braga, D., Maini, L., Grepioni, F. Mechanochemical preparation of co-crystals. Chem. Soc. Rev. 2013, 42, 7638–7648.

[9] Friščić, T., Jones, W. Recent Advances in Understanding the Mechanism of Cocrystal Formation via Grinding. Cryst. Growth Des. 2009, 9, 1621–1637.

[10] Friščić, T., Childs, S. L., Rizvi, S. A. A., Jones, W. The role of solvent in mechanochemical and sonochemical cocrystal formation: a solubility-based approach for predicting cocrystallisation outcome. Cryst. Eng. Comm. 2009, 11, 418–426.

[11] Co-crystals Preparation, Characterization and Applications. Aakeröy, C. B., Sinha, A. S., Eds. Royal Society of Chemistry, 2018.

[12] Delori, A., Friščić, T., Jones, W. The role of mechanochemistry and supramolecular design in the development of pharmaceutical materials. Cryst. Eng. Comm. 2012, 14, 2350–2362.

[13] Weyna, D. R., Shattock, T., Vishweshwar, P., Zaworotko, M. J. Synthesis and Structural Characterization of Cocrystals and Pharmaceutical Cocrystals: Mechanochemistry vs Slow Evaporation from Solution. Cryst. Growth Des. 2009, 9, 1106–1123.

[14] Karki, S., Fabian, L., Friščić, T., Jones, W. Powder X-ray Diffraction as an Emerging Method to Structurally Characterize Organic Solids. Org. Lett. 2007, 9, 3133–3136.

[15] Aakeröy, C. B., Desper, J., Smith, M. M. Constructing, deconstructing, and reconstructing ternary supermolecules. Chem. Commun. 2007, 3936–3938.

[16] Topić, F., Rissanen, K. Systematic Construction of Ternary Cocrystals by Orthogonal and Robust Hydrogen and Halogen Bonds. J. Am. Chem. Soc. 2016, 138, 6610–6616.

[17] Friščić, T., Trask, A. V., Jones, W., Motherwell, W. D. S. Screening for inclusion compounds and systematic construction of three-component solids by liquid-assisted grinding. Angew. Chem., Int. Ed. 2006, 45, 7546–7550.

[18] Shan, N., Toda, F., Jones, W. Mechanochemistry and co-crystal formation: effect of solvent on reaction kinetics. Chem. Commun. 2002, 2372–2373.

[19] Trask, A. V., Motherwell, W. D., Jones, W. Solvent-drop grinding: green polymorph control of cocrystallisation. Chem. Commun. 2004, 890–891.

[20] Braga, D., Maini, L., Giaffreda, S. L., Grepioni, F., Chierotti, M. R., Gobetto, R. Supramolecular Complexation of Alkali Cations through Mechanochemical Reactions between Crystalline Solids. Chem. Eur. J. 2004, 10, 3261–3269.

[21] Braga, D., Giaffreda, S. L., Grepioni, F., Pettersen, A., Maini, L., Curzi, M., Polito, M. Mechanochemical preparation of molecular and supramolecular organometallic materials and coordination networks. Dalton Trans. 2006, 1249–1263.

[22] Lukin, S., Stolar, T., Tireli, M., Blanco, M. V., Babić, D., Friščić, T., Užarević, K., Halasz, I. Tandem in situ monitoring for quantitative assessment of mechanochemical reactions involving structurally unknown phases. Chem. Eur. J. 2017, 23, 13941–13949.

[23] Gutmann, V. Solvent effects on reactivities of organometallic compounds. Coord. Chem. Rev. 1976, 18, 225–255.

[24] Taylor, C. R., Day, G. M. Evaluating the Energetic Driving Force for Co-crystal Formation. Cryst. Growth Des. 2018, 18, 892–904.

[25] Chan, H. C. S., Kendrick, J., Neumann, M. A., Leusen, F. J. J. Towards ab initio screening of co-crystal formation through lattice energy calculations and crystal structure prediction of nicotinamide, isonicotinamide, picolinamide and paracetamol multi-component crystals. Cryst. Eng. Comm. 2015, 13, 3799–3807.

[26] Bučar, D.-K., Day, G. M., Halasz, I., Zhang, G. G. Z., Sander, J. R. G., Reid, D. G., MacGillivray, L. R., Duer, M. J., Jones, W. The curious case of (caffeine)·(benzoic acid): how heteronuclear seeding allowed the formation of an elusive co-crystal. Chem. Sci. 2013, 3, 4417–4425.

[27] Bučar, D.-K., Lancaster, R. W., Bernstein, J. Disappearing Polymorphs Revisited. Angew. Chem. Int. Ed. 2015, 54, 6972–6993.

[28] Dunitz, J. D., Bernstein, J. Disappearing Polymorphs. Acc. Chem. Res. 1995, 28, 193–200.

[29] Belenguer, A. M., Lampronti, G. I., Cruz-Cabeza, A. J., Hunter, C. A., Sanders, J. K. M. Solvation and surface effects on polymorph stabilities at the nanoscale. Chem. Sci. 2016, 7, 6617–6627.

[30] Trask, A. V., Shan, N., Motherwell, W. D., Jones, W., Feng, S., Tan, R. B., Carpenter, K. J. Selective polymorph transformation via solvent-drop grinding. Chem. Commun. 2005, 7, 880–882.

[31] Hasa, D., Jones, W. Screening for new pharmaceutical solid forms using mechanochemistry: a practical guide. Adv. Drug Deliv. Rev. 2017, 117, 147–161.

[32] Lukin, S., Lončarić, I., Tireli, M., Stolar, T., Blanco, M. V., Lazić, P., Užarević, K., Halasz, I. Experimental and Theoretical Study of Selectivity in Mechanochemical Cocrystallization of Nicotinamide with Anthranilic and Salicylic Acid. Cryst. Growth Des. 2018, 18, 3, 1539–1547. Cryst. Growth Des. 2018, 18, 1539–1547.

[33] Ojala, W. H., Etter, M. C. Polymorphism in anthranilic acid: a reexamination of the phase transitions. J. Am. Chem. Soc. 1992, 114, 10288–10293.

[34] Stilinović, V., Horvat, G., Hrenar, T., Nemec, V., Cinčić, D. Halogen and Hydrogen Bonding between (N-Halogeno)-succinimides and Pyridine Derivatives in Solution, the Solid State and in Silico. Chem. Eur. J. 2017, 23, 5175–5182.

[35] Mavračić, J., Cinčić, D., Kaitner, B. Halogen bonding of N-bromosuccinimide by grinding. Cryst. Eng. Comm. 2016, 18, 3343–3346.

[36] Bolla, G., Nangia, A. Pharmaceutical co-crystals: walking the talk. Chem. Commun. 2016, 52, 8342–8360.

[37] Duggirala, N. K., Perry, M. L., Almarsson, Ö., Zaworotko, M. J. Pharmaceutical co-crystals: along the path to improved medicines. Chem. Commun. 2016, 52, 640–655.

[38] Kavanagh, O. N., Croker, D. M., Walker, G. M., Zaworotko, M. J. Pharmaceutical cocrystals: from serendipity to design to application. Drug Discov. Today 2019, 24, 796–804.

[39] Durgashankar, P., Ashish, P., Anjali, P. A review on co-crystal. J. Sci. Res. Pharm. 2012, 1, 26–28.

[40] Patel, P. V., Brahmbhatt, H., Upadhyay, U. M., Shah, V. A review on increased therapeutic efficiency of drugs by pharmaceutical co-crystal approach. Int. J. Pharm. Sci. Rev. Res. 2012, 16, 40–148.

[41] For Compounds 'Generally Recognized As Safe' (GRAS) by the Food and Drug Administration (FDA), please visit: https://www.fda.gov/food/food-ingredients-packaging/generally-recognized-safe-gras accessed May 15, 2020.

[42] Bailey, C. W., Bright, J. R., Jasper, J. J. A. Study of the Binary System Nicotinamide – Ascorbic Acid. J. Am. Chem. Soc. 1945, 67, 1184–1186.

[43] Stolar, T., Lukin, S., Tireli, M., Sović, I., Karadeniz, B., Kereković, I., Matijašić, G., Gretić, M., Katančić, Z., Dejanović, I., Di Michiel, M., Halasz, I., Užarević, K. Control of Pharmaceutical Cocrystal Polymorphism on Various Scales by Mechanochemistry: Transfer from the Laboratory Batch to the Large-Scale Extrusion Processing. ACS Sustainable Chem. Eng. 2019, 7, 7102–7110.

[44] Caira, M. R., Nassimbeni, L. R., Wildervanck, A. F. Selective formation of hydrogen bonded cocrystals between a sulfonamide and aromatic carboxylic acids in the solid state. J. Chem. Soc. Perkin Trans. 1995, 2, 2213–2216.

[45] Fleischman, S. G., Kuduva, S. S., McMahon, J. A., Moulton, B., Walsh, R. D. B., Rodríguez-Hornedo, N., Zaworotko, M. J. Crystal engineering of the composition of pharmaceutical phases: multiple-component crystalline solids involving carbamazepine. Cryst. Growth Des. 2003, 3, 909–919.

[46] Halasz, I., Puškarić, A., Kimber, S. A. J., Beldon, P. J., Belenguer, A. M., Adams, F., Honkimäki, V., Dinnebier, R. E., Patel, B., Jones, W., Štrukil, V., Friščić, T. Real-Time In Situ Powder X-ray Diffraction Monitoring of Mechanochemical Synthesis of Pharmaceutical Cocrystals. Angew. Chem. Int. Ed. 2013, 52, 11538–11541.

Guido Ennas, Alessandra Scano

6 Coordination polymers by a mechanochemical approach

This chapter deals with the use of mechanochemistry in the synthesis of coordination polymers (CPs). First, a general overview on these very interesting compounds is presented. Then, an assortment of reactions to prepare CPs by mechanosynthesis at the laboratory scale is described. To verify the success of the proposed syntheses, the powder X-ray diffraction (PXRD) technique is suggested. The CP structure patterns can be found in the Cambridge Structure Database.[1]

6.1 Coordination polymers and metal–organic frameworks: a brief overview

A CP is composed from metal centers (nodes) to which organic ligands (linkers) are coordinated and which bridge metal centers leading to coordination frameworks extending in 1, 2 or 3 dimensions.[2] Nodes could be ions or polynuclear clusters with high stability, the so-called secondary building units (SBU). Linkers contain variable length chains (spacer) that contribute to geometry of the structure changing the distance between the nodes (Figure 6.1).

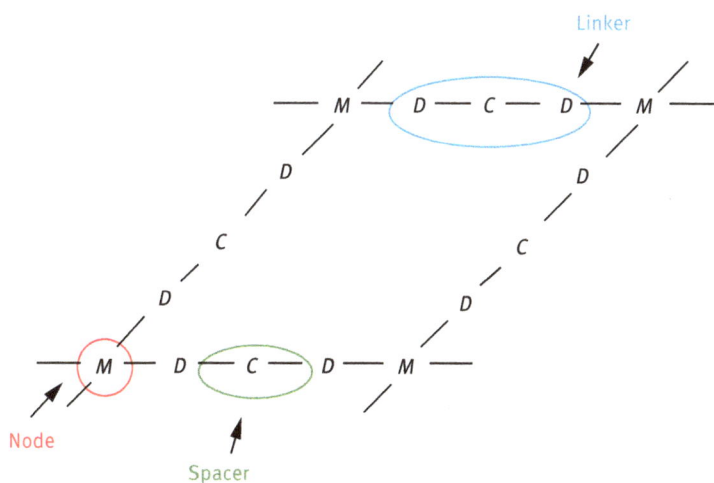

Figure 6.1: Schematic representation of a coordination polymer.

https://doi.org/10.1515/9783110608335-006

CPs constitute an interdisciplinary field, with its origins in solid state, inorganic and coordination chemistry, that has expanded rapidly in the last two decades, attracting interest of the chemical industry and crystal engineering.[3]

Metal–organic frameworks (MOFs) are a subset of CPs (Scheme 6.1), which contain voids in their 2D or 3D coordination frameworks and may exhibit high porosities.[2] Internal voids could reach even 90% of the total material volume, achieving high surface areas reaching up to 6000 m^2/g.[4]

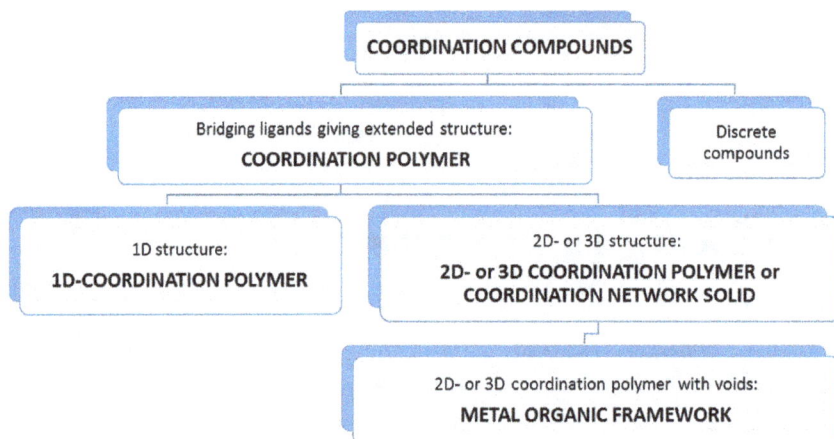

Scheme 6.1: A hierarchy relation between coordination compounds, coordination polymers and metal–organic frameworks.

6.2 Strategies for the synthesis of coordination polymers

Historical development of the field of CPs is an excellent example of interdisciplinary research, developed by the effort of two different scientific backgrounds. The first one is represented by crystal engineering and coordination chemistry that have their principles focused on the assembly of organic and inorganic building blocks in order to form porous structures. The second one is the chemistry of zeolites that started to look at the use of organic molecules, not only as structure-directing agents, but also as reactants to be incorporated in the framework structure. These explain the variety of synthetic methodologies and strategies nowadays adopted in the field of CP synthesis.

The conventional method to obtain CPs is the solvo/hydrothermal synthesis; some alternatives routes include microwave-assisted, sonochemical, electrochemical and mechanochemical methods.[5]

6.3 Mechanochemical reactivity leading to coordination bonds

CPs can be constructed using different mechanochemical methodologies discussed in the previous chapters, such as solvent-free neat grinding (NG) and liquid-assisted grinding (LAG) synthetic strategies, in order to achieve different types of reactions that lead to formation of coordination bonds. Five different reaction types for the construction of CPs are available: (1) ligand addition, (2) ligand exchange, (3) acid–base reactions, (4) dehydration reactions and (5) oxidative addition.[5]

Some of the earlier listed processes can be coupled as one-pot pathways. Mechanochemistry is an ideal approach to these multicomponent reactions.

6.3.1 Ligand addition

One of the simplest methodologies to CP construction via mechanochemistry is the addition of neutral ligands, where the addition of a multitopic organic ligand to the metal center leads to the formation of a new discrete (1D) or extended (2D or 3D) structure. A typical stepwise mechanism of the ligand addition has been observed during manual NG of $ZnCl_2$ with the diamine [2.2.2]-diazabicyclooctane (dabco).[6] The first step is the formation of an intermediate hydrate $ZnCl_2(dabco)\cdot 4H_2O$ phase, which upon heating or further grinding led to the 1D zigzag anhydrous polymer $ZnCl_2(dabco)$. The same reaction carried out in dry air and with nonhydrated reactants can lead directly to the $ZnCl_2(dabco)$ polymer.[7]

An example of this reaction type by milling apparatus was reported by Pichon and James using solvent-free mechanosynthesis for the preparation of the $Cu(acac)_2$ $(bipy)_n$ 1D polymer (Figure 6.2), starting from copper(II) acetylacetonate and an excess of 4,4'-bipyridine (bipy)[8] (eq. (6.1)):

$$2\,Cu(acac)_{2\,(s)} + 3\,bipy_{(s)} \xrightarrow{\;\;\otimes\!\!\otimes\;\;} 2[Cu(acac)_2 bipy]_{(s)} + bipy_{(s)}$$

Equation 6.1. Reaction conditions: $Cu(acac)_2$ 1.0 mmol, bipy 1.5 mmol.

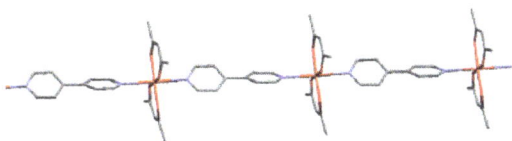

Figure 6.2: Structure of the $[Cu(acac)_2(bipy)]_n$ polymer according to the CSD entry DEDJAX. Reproduced from reference[8] (with permission from the Royal Society of Chemistry).

The obtained CP has a linear structure based on pseudooctahedral copper centers, with equatorial acetylacetonate ligands and axial bipy linkers.[8] The remaining bipy in the reaction is consistent with the obtained CP stoichiometry (1:1) being different to that used in the reaction and can be removed by washing. The related milling conditions are reported in the Table 6.1. The milling speed, milling time, milling-rest period and reaction atmosphere are not reported in the original work[8]; therefore, we suggest 25–30 Hz, up to 60 min, without rest period and using air atmosphere, respectively.

Table 6.1: Process parameters for the synthesis of [Cu(acac)$_2$(bipy)]$_n$ by mechanochemistry.

Parameters of the milling process	Method
Type of mill	Mixer mill
Jar volume (mL)	20
Jar material	SS[a]
Number of balls	1
Balls diameter (mm)	10
Weight of each ball (g)	4.1
Milling speed	n.r.[b] (25–30 Hz)[c]
Milling time (min)	n.r.[b] (up to 60)[c]
Milling and resting period (min)	n.r.[b] (none)[c]
Grinding additive	–
Reaction atmosphere	n.r.[b] (air)[c]
Reaction scale (mmol)	1.0

[a]SS, stainless steel;
[b]n.r., not reported;
[c]suggested conditions.

The CP preparation by ligand addition can also be carried out according to the LAG method. Addition of a liquid phase is useful to accelerate or even enable a mechanochemical reaction, as well as to provide an opportunity for molecular inclusion in the nascent host CP. An example is the polymer obtained from CuCl$_2$ and 1,4-diaminocyclohexane (dace) with inclusion of DMSO, which gives rise to CuCl$_2$(dace)$_n$·DMSO. The analogous coordination polymer is not obtained in solvent-free grinding.[9]

6.3.2 Ligand exchange

Another reaction type in mechanosynthesis of CPs is the ligand exchange, which involves the mechanochemical replacement of ligands on a metal center. The mechanochemical formation of new metal–ligand bonds often involves breaking connections between metal atoms and water, followed by water removal. Manual grinding of silver acetate and a diamine 1,4-diazabicyclo[2.2.2]octane (dabco) re-sults in displacement of the acetate ligands with dabco along with immediate water absorption from air that gives rise to the formation of $Ag(OAc)(dabco)_2 \cdot 5H_2O$ (eq. (6.2) and Figure 6.3).[7] This demonstrates the importance of the atmosphere for mechanosynthesis:

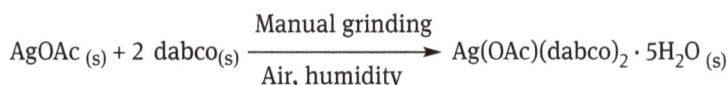

$$AgOAc_{(s)} + 2\ dabco_{(s)} \xrightarrow[\text{Air, humidity}]{\text{Manual grinding}} Ag(OAc)(dabco)_2 \cdot 5H_2O_{(s)}$$

Equation 6.2. Preparation of the coordination network for $Ag(OAc)(dabco)_2 \cdot 5H_2O$ by manual grinding: 5 min using an agate mortar and pestle. Reaction conditions for method: $AgOAc_{(s)}$ (1.5 mmol), $dabco_{(s)}$ (3.1 mmol), $OAc = CH_3COO^-$; dabco = 1,4-diazabicyclo[2.2.2]octane.

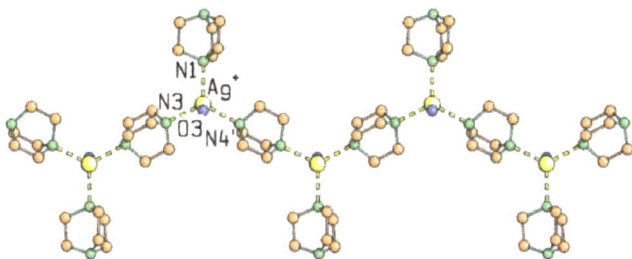

Figure 6.3: Structure of the $[Ag(OAc)(dabco)_2 \cdot 5H_2O]$ compound (for clarity, water was omitted). Reproduced from reference[7] (with permission from the Royal Society of Chemistry).

6.3.3 Acid–base reactions

CP synthesis based on acid–base reactions involves proton transfer where the forma-tion of a new metal–organic structure results from the protonation and expulsion of a ligand attached to a metal precursor. The latter could be an acetate, chloride, car-bonate, hydroxide or an oxide. An example of the synthesis of CP by acid–base reac-tions is the preparation of $\{[X(o\text{-}ABA)_2(H_2O)_3]\}_n$ by milling a 2:1 ratio of anthranilic acid (o-aminobenzoic acid, o-ABAH) together with $X(OH)_2$ (X = Ca, Sr or Ba) in a plan-etary mill[10] (eq. (6.3)):

$$\text{Ca(OH)}_{2\,(s)} + 2\,o\text{-ABAH}_{(s)} \xrightarrow[\text{LAG (H}_2\text{O, 130 μL)}]{} \{[\text{Ca}(o\text{-ABA})_2(\text{H}_2\text{O})_3]\}_{n\,(s)}$$

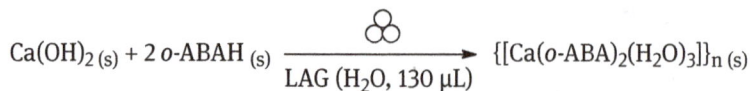

Equation 6.3. Preparation of $\{[\text{Ca}(o\text{-ABA})_2(\text{H}_2\text{O})_3]\}_n$ in a planetary mill. Reaction conditions for method: Ca(OH)_2 2.9 mmol, o-ABAH 5.8 mmol.

130 μL of water were added to the powder mixture to improve the crystallinity of the final product and reduce the milling time from 4 h to 1 h. In the absence of water, a product with lower crystallinity was formed, which can be estimated by comparing peak widths in their powder diffraction patterns. The structure of $\{[\text{Ca}(o\text{-ABA})_2(\text{H}_2\text{O})_3]\}_n$ is shown in Figure 6.4, while the milling parameters are given in Table 6.2.

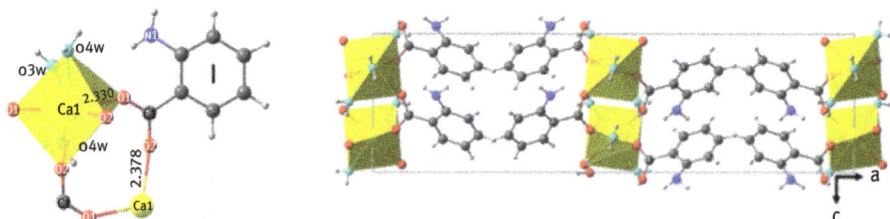

Figure 6.4: The structure of the $\{[\text{Ca}(o\text{-ABA})_2(\text{H}_2\text{O})_3]\}_n$ coordination polymer obtained by acid–base reaction.[10] Reproduced from reference[10] (with permission from the Royal Society of Chemistry).

Table 6.2: Process parameters for the synthesis of $\{[\text{Ca}(o\text{-ABA})_2(\text{H}_2\text{O})_3]\}_n$ by mechanochemistry.

Key parameters of the milling process	Method
Type of mill	Planetary mill
Jar volume (mL)	45
Jar material	Si_3N_4
Number of balls	5
Balls diameter (mm)	12
Weight of each ball (g)	2.8
Milling speed	600 rpm
Milling time (min)	60
Milling and resting period (min)	n.r.[a]
Grinding additive	130 μL H_2O
Reaction atmosphere	Air
Reaction scale (mmol)	2.9

[a]n.r., not reported.

In this context, it is also possible to use poorly soluble metal oxides; these are very attractive for their low cost, availability and because water is their only byproduct. Several CPs have been constructed by using zinc oxide with different organic ligands. An example of CPs with metal oxides is shown by Friščić and coworkers.[11] Different products were obtained from zinc oxide and fumaric acid (fum) by using various solvents for LAG synthesis. Anhydrous 3D nonporous coordination polymer of zinc fumarate was obtained when using methanol and ethanol, while a dihydrate form, which is a 2D polymer, was obtained when using a 1:1 mixture of water and ethanol. Moreover, the same reaction using increasing amounts of water revealed two different 1D structures, a tetrahydrate and a pentahydrate. It is noteworthy that in this example LAG syntheses result in four different structures by changing only the liquid phase or its amount (Figure 6.5).

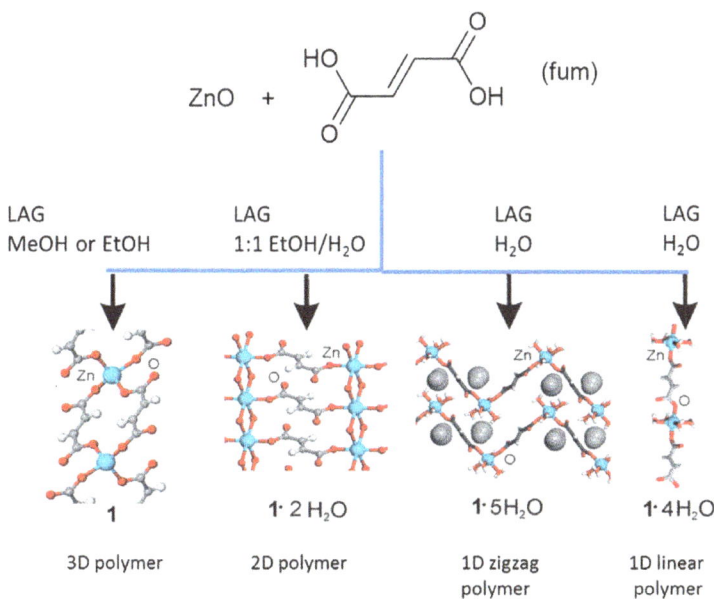

Figure 6.5: Comparison of different products obtained from zinc oxide and fumaric acid (fum) by using various solvents for LAG syntheses. Guest water molecules are shown in gray. Reproduced from reference[11] (with permission from the Royal Society of Chemistry).

6.3.4 Dehydration reactions

Synthesis of CPs by dehydration reactions occurs when a metal center loses coordinated water resulting in a change of the structure. This is the case of a mechanochemical dehydration and polymerization of small molecule complexes into MOFs in a LAG approach, using a liquid with high affinity for water. Such LAG dehydration is readily reversed by milling the obtained MOFs with water. Such mechanochemical dehydration is more effective than heating or immersion in bulk solvents.

A fascinating example is reported by Wang and coworkers, who prepared [Zn(INA)$_2$ (H$_2$O)$_4$] and [Zn(INA)$_2$] starting from isonicotinic acid (HINA) and ZnO in the 2:1 stoichiometric ratio and in the presence of a small quantity of water or methanol, respectively.[12] Using water, a discrete complex [Zn(INA)$_2$(H$_2$O)$_4$] was obtained within 30 min of milling, while the use of methanol leads to a 3D triply-interpenetrated diamondoid network [Zn(INA)$_2$] in which the Zn centers are pseudo-tetrahedral (Figure 6.6).

[Zn(INA)$_2$(H$_2$O)$_4$]   [Zn(INA)$_2$]

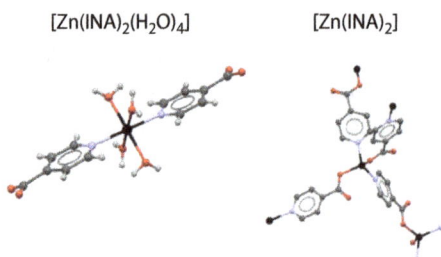

Figure 6.6: Structures of the discrete complex Zn(INA)$_2$(H$_2$O)$_4$ and of the diamondoid network Zn(INA)$_2$.

Moreover, clean interconversions (dehydration and rehydration reactions) between the discrete and polymeric forms are also possible by LAG with MeOH, during 90 min milling [Zn(INA)$_2$(H$_2$O)$_4$] loses water and is converted into Zn(INA)$_2$ and the reverse reaction occurs quantitatively by LAG with H$_2$O for 30 min. The same results are obtained in a conventional approach by prolonged immersion in the bulk solvents: requiring 3 days in methanol to dehydrate and 6 h in water to rehydrate. The overall reaction scheme is shown in eq. (6.4) while the mechanochemical process parameters for the synthesis of Zn(INA)$_2$ and its interconversion to Zn(INA)$_2$(H$_2$O)$_4$ are given in Tables 6.3 and 6.4.

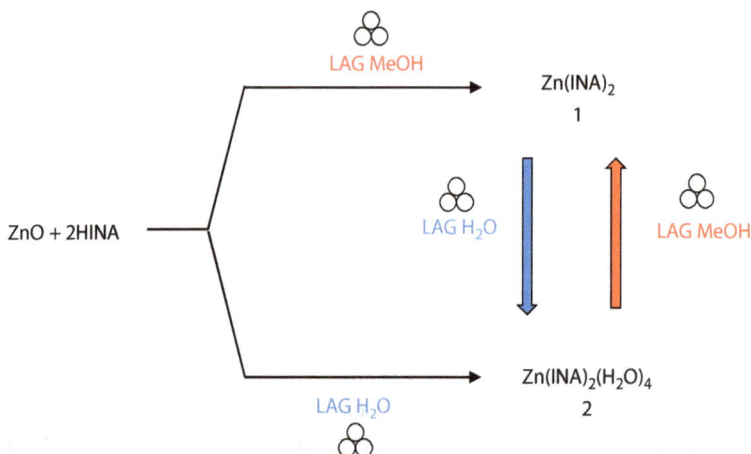

Equation 6.4. [Zn(INA)$_2$(H$_2$O)$_4$] and [Zn(INA)$_2$] by mechanochemistry. Reaction conditions for method: ZnO 1 mmol, HINA 2 mmol.

Table 6.3: Process parameters for the synthesis by mechanochemistry of $Zn(INA)_2(H_2O)_4$ in the presence of small quantity of water, or Zn $(INA)_2$ in the presence of a small amount of methanol.

Key parameters of the milling process	Method
Type of mill	Planetary mill
Jar volume (mL)	50
Jar material	SS[a]
Number of balls	15
Balls diameter (mm)	5
Weight of each ball (g)	n.r.[b]
Milling speed	650 rpm
Milling time (min)	30
Milling and resting period (min)	n.r.[b]
Grinding additive	H_2O or MeOH 100µL
Reaction atmosphere	n.r.[b]
Reaction scale (mmol)	1

[a]SS, stainless steel;
[b]n.r., not reported.

6.3.5 Oxidative addition

One of the fundamental reactions of the organometallic chemistry is represented by the oxidative addition. During this reaction, the increase of the coordination number of a central metal atom along with its oxidation state is achieved. In particular, this type of reactions is currently gaining interest in the context of metal–organic materials prepared by multicomponent, *one-pot* reactions.

A noteworthy example is reported by Hernandez and coworkers, who describe a combination of oxidative addition and ligand exchange reactions to allow for the mechanochemical *one-pot* synthesis of a complex organometallic fluoride complex from altogether five components. The metal carbonyl is used as the simplest and most readily accessible metal precursor.[13] For a deeper insight in the process parameters related to the preparation of this material, the reader is invited to refer to the original work.[13]

Table 6.4: Process parameters for the Interconversion reactions by LAG of $[Zn(INA)_2(H_2O)_4]$ **2** into $[Zn(INA)_2]$ **1** and vice versa.

Key parameters of the milling process	Method
Type of mill	Planetary mill
Jar volume (mL)	50
Jar material	SS[a]
Number of balls	15
Balls diameter (mm)	5
Weight of each ball (g)	n.r.[b]
Milling speed	650 rpm
Milling time (min)	90[c] or 30[d]
Milling and resting period (min)	n.r.[b]
Grinding additive	MeOH 200 µL[c] or H_2O 100 µL[d]
Reaction atmosphere	n.r.[b]
Reaction scale (mmol)	1

[a]SS, stainless steel;
[b]n.r., not reported;
[c]for dehydration reaction;
[d]for rehydration reaction.

Conclusion

Synthesis of CPs represents one of the most recent and exciting branches of mechanochemistry. The latter has demonstrated the capability to construct complex metal–organic materials starting from simple components and profiting of hierarchical self-assembly processes. Moreover, the use of mechanochemistry for the preparation of materials whose synthetic approach is dominated by solution methods is a noteworthy achievement.

We hope this chapter could shed some light on the use of mechanochemistry for the development of CPs. Still a lot needs to be done and the integration of mechanosynthesis with reliable methods for characterizing both reaction intermediates and final products will open new opportunities and horizons in the field.

References

[1]   Groom, C. R., Bruno, I. J., Lightfoot, M. P., Ward, S. C. The Cambridge Structural Database Acta Cryst. 2016, B72, 171–179. For more information on The Cambridge Structural Database (CSD) see also: https://www.ccdc.cam.ac.uk/solutions/csd-system/components/csd/.

[2]   Batten, S. R., Champness, N. R., Chen, X. M., Garcia-Martinez, J., Kitagawa, S., Ohrstrom, L., O'Keefe, M., Suh, M. P., Reedijk, J. Terminology of metal-organic frameworks and coordination polymers (IUPAC recommendations 2013). Pure Appl. Chem. 2013, 85, 1715–1724.

[3]   Yaghi, O. M., Li, H., Davis, C., Richardson, D., Groy, T. L. Synthetic strategies, structure patterns, and emerging properties in the chemistry of modular porous solids. Acc. Chem. Res. 1998, 31, 474–484.

[4]   Zhou, H.-C., Long, J. R., Yaghi, O. M. Introduction to Metal–Organic Frameworks. Chem. Rev. 2012, 112, 673–674.

[5]   Tong, M.-L., Chen, X.-M. Synthesis of Coordination Compounds and Coordination Polymers. In: Modern Inorganic Synthetic Chemistry (2nd Edition), Xu, R., Xu, Y., Eds. Elsevier, 2017, Ch. 8, 189–217. ISBN: 978–0–444–63591–4.

[6]   Friščić, T. Ball-milling Mechanochemical Synthesis of Coordination Bonds: Discrete Units, Polymers and Porous Materials. In: Ball Milling Towards Green Synthesis: Applications, Projects, Challenges, Ranu, B., Stolle, A., Ed. Royal Society of Chemistry, 2015, 151–189.

[7]   Braga, D., Giaffreda, S. L., Grepioni, F., Polito, M. Mechanochemical and solution preparation of the coordination polymers Ag [N(CH$_2$CH$_2$)$_3$N]$_2$ [CH$_3$COO]·5 H$_2$O and Zn [N(CH$_2$CH$_2$)$_3$N]Cl$_2$. Crys. Eng. Comm. 2004, 6, 459–462.

[8]   Pichon, A., James, S. L. An array-based study of reactivity under solvent-free mechanochemical conditions – insights and trends. Crys. Eng. Comm. 2008, 10, 1839–1847.

[9]   Braga, D., Curzi, M., Johansson, A., Polito, M., Rubini, K., Grepioni, F. Simple and quantitative mechanochemical preparation of a porous crystalline material based on a 1D coordination network for uptake of small molecules. Angew. Chemie Int. Ed. 2006, 45, 142–146.

[10] Al-Terkawi, A.-A., Scholz, G., Prinz, C., Emmerling, F., Kemnitz, E. Ca-, Sr-, and Ba-Coordination polymers based on anthranilic acid via mechanochemistry. Dalton Trans. 2019, 48, 6513–6521.

[11] Friščić, T., Fábián, L. Mechanochemical conversion of a metal oxide into coordination polymers and porous frameworks using liquid-assisted grinding (LAG). Crys. Eng. Comm. 2009, 11, 743–745.

[12] Wang, P., Li, G., Chen, S., James, S. L., Yuan, W. Mechanochemical interconversion between discrete complexes and coordination networks – formal hydration/dehydration by LAG. Crys. Eng. Comm. 2012, 14, 1994–1997.

[13] Hernandez, J. G., Butler, I. S., Friščić, T. Multi-step and multi-component organometallic synthesis in one pot using orthogonal mechanochemical reactions. Chem. Sci. 2014, 5, 3576–3582.

Index

active pharmaceutical ingredients (APIs) 4, 13, 41, 62, 79
acyl transfer agent 60
alloys 23, 25, 34
amides
– CDI-mediated 60
amorphization 11, 17, 23, 25, 26, 82
arylboronic *See* organoboron
attritors 11
azides 47
– in situ preparation 47

ball-mill
– automated 11
benign by design 41
Biginelli reaction 58
binary alloys 26
– with boron or silicon 26
bismuth subsalicylate 63

carbamates
– CDI-mediated 61
carbonyldiimidazole (CDI) 60
C–C bond formation *See* cross-coupling organoboron
ceramic nanocomposites *See* nanocomposites 34
characterisation
– powder X-ray diffraction 15, 23, 91
cinnabar 3
clean manufacturing 4, 9, 41, 62
click reaction 47
cocrystal 5, 71
– halogen bonding 78
– hydrogen-bonded 72
– in situ anaysis 82
– pharmaceutical 79
– proton transfer 74
– screening 71
– selectivity 75
– solubility 71
– stoichiometry 72, 78
– supramolecular interactions 71
– vitamin 80
– three-component 71
composite 34
contamination

– metal leaching 14
coordination polymer
– acid–base 95
– by dehydration 97
– ligand addition 93
– ligand exchange 95
– metal–organic frameworks 92
– oxidative addition 99
– secondary building units 91
– synthesis 92
coordination polymer (CP) 91
copper-catalyzed cycloaddition *See* click reaction
cross-coupling
– Mizoroki-Heck 44
– Negishi 48
– palladium-catalyzed 42
– sacrificial base 43
– Sonogashira 46
– Suzuki–Miyaura 42

dantrolene 63
diasteroselectivity 54
Diels–Alder
– hetero- 54
Diels–Alder reaction 54
diene *See* Diels–Alder
dienophile *See* Diels–Alder
dry milling *See* neat grinding 14

economy
– energy 9, 62
– reagent 54, 62, 64
– atom 47, 63
– waste 47
– time 9, 62
elemental substances 23
eta (η) parameter 14, 72
ex situ analysis 15, 78

F. M. Flavitsky 3
Faraday 3 ,23

generally recognized as safe (GRAS) 79
Glaser reaction *See* click reaction
gram scale *See* mechanochemistry 11

https://doi.org/10.1515/9783110608335-007

www.ingramcontent.com/pod-product-compliance
Lightning Source LLC
Chambersburg PA
CBHW081548220326
41598CB00036B/6606